# SHARING OUR INTELLECTUAL TRACES
## Narrative Reflections from Administrators of Professional, Technical, and Scientific Communication Programs

Tracy Bridgeford
*University of Nebraska at Omaha*

Karla Saari Kitalong
*Michigan Technological University*

Bill Williamson
*Saginaw Valley State University*

Baywood's Technical Communications Series
*Series Editor: Charles H. Sides*

Routledge
Taylor & Francis Group
LONDON AND NEW YORK

First published 2014 by Baywood Publishing Company, Inc.

Published 2017 by Routledge
2 Park Square, Milton Park, Abingdon, Oxon OX14 4RN
711 Third Avenue, New York, NY 10017, USA

*Routledge is an imprint of the Taylor & Francis Group, an informa business*



Library of Congress Catalog Number: 2013034714
ISBN 13: 978-0-89503-869-2 (hbk)
ISBN 13: 978-0-89503-870-8 (pbk)

**Library of Congress Cataloging-in-Publication Data**

Bridgeford, Tracy, 1960–
Sharing our intellectual traces : narrative reflections from administrators of professional, technical, and scientific communication programs / Tracy Bridgeford, University of Nebraska at Omaha, Karla Saari Kitalong, Michigan Technological University, Bill Williamson, Saginaw Valley State University.
pages cm. -- (Baywood's technical communications series)
Includes bibliographical references and index.
ISBN 978-0-89503-869-2 (cloth : alk. paper) -- ISBN 978-0-89503-870-8 (pbk.) -- ISBN 978-0-89503-871-5 (e-pub) -- ISBN 978-0-89503-872-2 (e-pdf) 1. Communication of technical information--Management. 2. Technology--Information services--Management. 3. Science--Information services--Management. 4. Information resources management. I. Kitalong, Karla Saari, 1952– II. Williamson, Bill, 1968– III. Title.
T10.5.B73 2013
601.4--dc23

2013034714

# Table of Contents

FOREWORD

# Understanding the Power of Narrative in Shaping a Field

***Kirk St.Amant***

It could be said that humans are a narrative species. That is, of all of the creatures in nature, humans seem to be unique in creating stories in order to make meaning out of their experiences, to understand events, or to share their interpretations with others. In fact, one could claim humans alone use narratives to pass knowledge on from generation to generation. After all, the narrative form of the fairy tale is one of the first mechanisms we use to teach children what constitutes "good" and "bad" behavior as well as what rewards are bestowed upon the good versus what punishments await the bad.

Narratives are also very powerful rhetorical devices. By creating a context for understanding events, the stories we tell and how we tell them shape our understandings of what took place at various points in time. The descriptors we use when telling stories, for example, allow us to control how an audience perceives the actions taken by individuals in relation to different events. Similarly, the way in which a narrative unfolds allows the storyteller to control the sequence in which events took place and to create a sense of causality in terms of what prior actions resulted in later effects. In these ways, truly effective stories that recount past actions can actually eclipse the facts of the events themselves. And such processes can spread until the legend of what took place becomes reality in the minds of a greater group of individuals. At that point, the perspective conveyed by the narrative constitutes what is accepted as "true" by the related group. As the 1962 film *The Man Who Shot Liberty Valance* puts it, "When the legend becomes the truth, print the legend."

Perhaps one of the most interesting aspects of the use of narrative as a rhetorical device is how stories—or an appeal to narrative, if you will—can create conversation among individuals. Phrases such as "Do I have a story for you." or "I have an interesting story about that." or "Would you like to hear a story about how . . . ?" inevitably lead us to invite the speaker to share that story with us. Moreover, such devices also often prompt us to share our own stories related to that topic or theme. Through such narrative-based conversations, groups of individuals often begin to create a greater, shared story that can become a framework for mutual understanding and for future, common interaction. In this way, one could say that by creating such commonalities, narratives often serve as a foundation for culture.

At its core, culture is about commonality. That is, the members of a given culture generally share a common set of beliefs, values, attitudes, and behaviors that bind them together. Such commonalities are often connected to and conveyed through a language that helps the members of a culture discuss what they consider important to that culture. Within this context, narratives can be particularly powerful.

In many cases, cultures have a creation story—or narrative—that states how that culture came into being. Such narratives often define who the members of that culture are, what values the members of that culture hold dear, and what behaviors establish a foundation for evaluating right and wrong among the members of that culture. Similarly, the legends shared by the members of a culture further cement bonds of commonality by providing examples of what values and behaviors are considered "good" within that culture (e.g., the actions of individuals labeled as "heroes") and what behaviors are considered "bad" within the culture (e.g., the actions of persons identified as "villains"). These legends also provide the members of a culture with a lens through which they can view, interpret, and understand historical events based on how the members of that culture (and those from outside of the culture) are depicted in the recounting of past activities. Thus, narratives become an important, if not a central, way through which history is presented to and remembered by a greater group.

As with any culture, disciplines are also deeply connected to narratives. In such cases, the creation narrative of the culture often becomes stories about the formation and early history of a given field, the development of a particular program, or the recounting of events that shaped both. Similarly, narratives help the members of a disciplinary culture develop the vocabulary of that culture through stories of what was successful and what was not. And narratives also serve as effective tools for engendering discussion across a given field. Consider, for example, how many of us use conferences, listservs, or other professional forums to present stories of our experiences as well as invite others to share their own stories about what took place at their schools or within their programs. Again, through such narrative-based activities, commonality across a group can emerge, and disciplinewide mechanisms for understanding and addressing events can develop. For these reasons, all disciplines need to collect and to share their narratives.

As a relatively new and growing discipline, technical and scientific communication abounds with important and interesting narratives that have a great deal to offer members of the field. These range from stories of the individual faculty member/program administrator navigating her or his way through a greater department (usually an English department) to persons recounting how discipline-specific programs emerged out of various concentrations, degrees, or departments to become independent academic units/entities. Such narratives provide a rich set of experiences others can use as heuristics for contextualizing their own experiences or as approaches for addressing a wide range of events. Moreover, when collected in one place, this aggregation of narratives can serve as a critical resource and reference for other members of the field. I believe this edited collection is such a text.

Although it is impossible to collect and to share every narrative within a given field, works that bring together various narratives representing common themes regularly encountered within a field can provide a framework for understanding and addressing other, similar events. The editors of this collection have, in turn, assembled a group of authors who represent a broad range of experiences that underlie many of the core issues at play in our field today. These stories cover topics such as approaches to understanding assessment and developing educational outcomes. They also present examples of transformational experiences that have led to new ideas to and approaches for leadership as well as models and methods for interacting with the greater communities of which we are all a part. In so doing, the contributors share stories of obstacles they have encountered and approaches they have used to address forces that can affect individual members of our field as well as overall programs in our discipline.

Of equal importance is the range of individuals who share their stories in this text. These authors include faculty who have been actively involved in the field for 30+ years as well as newly minted PhDs. They represent both persons who have worked at numerous universities over the course of their careers and individuals who have dedicated a lifetime of experience to one institution. These contributors also provide perspectives that span the spectrum of higher education from research-intensive universities with PhD programs to more teaching-oriented institutions that focus exclusively on undergraduate education and service learning. Included in this mix are persons who have experiences related to on-site and online courses and degrees as well as representatives from key professional organizations such as the Council for Programs in Technical and Scientific Communication (CPTSC), the Society for Technical Communication (STC), the IEEE Professional Communication Society (IEEE PCS), the Association of Teachers of Technical Writing (ATTW), and many others. Thus, while no collection on this topic can ever be truly comprehensive in the stories it includes, this anthology represents an amazingly broad cross section of our field and the kinds of situations and concerns commonly encountered by its members. The result is a powerful framework and a related set of heuristics from which we can begin to understand those forces and factors that bind us together as a discipline.

It could be said that the story of every major undertaking begins with a single action, or step. In terms of collecting and sharing the narratives of our field, this volume represents such a first step. And, as with all narratives, this text also stands as an invitation for others to collect and share both their own stories and the narratives of others so the greater tale of the field might be explored in meaningful and important ways. For this reason, I encourage readers to view this text as an opportunity to engage in the greater narrative of the field and to collect and share experiences with others. After all, each of us has a story to tell.

CHAPTER 1

# Introduction: We Are the Stories We Tell

***Tracy Bridgeford, Karla Saari Kitalong, and Bill Williamson***

This edited collection serves two core functions for the professional, technical, and scientific communication (PTSC) community: to present a broad-spectrum examination of the intellectual and institutional challenges faced by administrators of academic programs related to the community and to construct for new and experienced administrators a context within which they might define or redefine their roles as program leaders. This collection will present readers with narrative reflections by members of the PTSC community who have served or who now serve as program administrators.

Although the collection directly engages administrative professionals in the target community, *Sharing Our Intellectual Traces* offers insight to other stakeholders as well, both within the core community and within similar or connected communities:

- Scholarly and teaching professionals who by design or necessity have recently adopted administrative roles and responsibilities,
- Graduate students engaged in professional development that includes program administration,
- Colleagues from similar or related disciplines (e.g., composition, rhetoric, communication) who share similar concerns, and
- Members of nonacademic professional communities who collaborate with PTSC programs and administrators.

As in similar books for professionals and scholars in composition and rhetoric, this collection is grounded in practice—the day-to-day concerns and activities of program administrators. However, unlike other books, this collection also emphasizes the need for tracing the intellectual debts of the authors to others who have impacted and informed their work as administrators. The PTSC community does not consistently and effectively share its reflections on the conceptual foundations for its work; the community has contented itself with stories of how it does work rather than demanding careful, disciplined examinations of the forces that influence that work. By asking authors to emphasize theory as they share their reflections on core administrative processes and significant moments in the histories of their associated programs, we hope to move this volume toward critical examinations of such intellectual traces. In effect, we seek to recast administration as a rich, complex intellectual engagement rather than an activity based only in institutional details and processes.

## SERVING ACADEMIC STAKEHOLDERS FOR *SHARING OUR INTELLECTUAL TRACES*

Although several scholarly venues are available in which the scholarship of technical and scientific communication program administration might be explored, the reality is that the literature of this field is disproportionately small given that administration is a significant element among the professional responsibilities of many PTSC faculty. The majority of PTSC faculty postings listed on CPTSC.org in 2004, 2005, and 2006 listed administrative work among the expectations and responsibilities of the position (70% of the 2004 postings, 80% in 2005, and 57% in 2006). These technical communication program directors, some of whom are entry-level assistant professors, face the same diverse challenges as writing program administrators (WPAs) and other program administrators; these include shrinking budgets, legislative mandates to demonstrate accountability, and the effects of a changing global marketplace on faculty recruitment, enrollments, and student job placement rates. Yet helpful articles to inform and assist them are in short supply.

Discussions about technical communication program administration are conspicuously absent from books about writing program administration (e.g., L'Eplattenier & Mastrangelo, 2004; McGee & Handa, 2005; Rose & Weiser, 1999, 2002). *Writing Program Administration: Journal of the Council of Writing Program Administrators*, a journal devoted to the scholarship of administration in writing studies, has not, to the best of our knowledge, ever addressed topics, issues, or problems related to technical communication. WPA's website[1] lists links to rhetoric and composition journals, but journals related to technical

[1] See http://wpacouncil.org/journal/index.html

communication are absent from this list, even though many technical communication faculty identify with the discipline as scholars. And, although the *Bulletin*, a publication of the Association of Departments of English (ADE), published frequently about technical communication in the 1980s, those articles have ceased in recent years, as technical communication has become a more naturalized part of English departments.

The notable edited collections in technical communication, such as *Technical Writing: Theory and Practice* (Fearing & Sparrow, 1989), *Foundations for Teaching Technical Communication: Theory, Practice, and Program Design* (Staples & Ornatowski, 1997), and *Computers and Technical Communication* (Selber, 1997), exert significant influence over classroom practice and curriculum design and have undoubtedly helped some technical communication program administrators set a direction for program development. But none of these important edited collections devotes more than a few chapters to a direct and explicit discussion of programmatic issues. Although important to program administrators, then, these scholarly works do not directly facilitate or even document the extensive programmatic work that occurs in technical communication departments.

Similarly, technical communication journals such as *Technical Communication Quarterly* (TCQ), the *Journal of Business and Technical Communication* (JBTC), and the *Journal of Technical Writing and Communication* (JTWC), with a few exceptions (e.g., Faber, 2002; Selber, 2004; Selber & Johnson-Eilola, 2001), neglect issues that theorize or otherwise directly affect program administration. Most of our discipline's established body of work focuses on teaching approaches such as service learning (e.g., Dubinsky, 2004; Matthews & Zimmerman, 1999,), research methods such as activity theory (e.g., Spinuzzi, 1996), historical studies (e.g., Brockmann, 2002; Tebeaux, 1997), or curricular topics such as engineering writing (e.g., Winsor, 1996). An exception to this rule is assessment; in this case alone, program administrators have benefited greatly from colleagues' expertise, often gained through lengthy experimentation as well as trial and error. However, assessment is only one among many administrative challenges faced by our colleagues in their programmatic roles.

Whether these absences are due to a lack of article submissions or a conscious decision on the part of the editors, the outcome is the same. Technical communication program administration is an undertheorized and marginalized practice. *Programmatic Perspectives*, the new journal of the Council for Programs in Technical and Scientific Communication (CPTSC), launched March 1, 2009, is an attempt to rectify this omission. The journal establishes technical communication program administration as a *bona fide* research area and sets the stage for further scholarship in that area.

## THE COLLECTION: SHARING INTELLECTUAL TRACES THROUGH CRITICAL NARRATIVES

First-person narratives about administering technical communication programs are among the types of publications missing from the literature. Such stories, part of the narrative turn in scholarship (Barry & Elmes, 1997; Blyler, 1996; Blyler & Perkins, 1999; Kreiswirth, 2000), are beginning to emerge in rhetoric and composition program administration (Dew & Horning, 2007; Enos & Borrowman, 2008; George, 1999), providing a complex context for understanding the everyday life of the administrator and program director. Because many technical communication tenure-track hires are tasked with creating and maintaining programs, knowledge about program development and administration is vital to their and their programs' success. Although program administration internships, coursework, and research assistantships such as the three of us enjoyed as graduate students afford some opportunities to see and participate in the day-to-day intricacies of program administration, participation is necessary for real understanding followed by action. Most of this in-depth experience happens, for graduate assistants and junior WPAs, in established programs that have already worked out some of the kinks. But tricky yet legitimate situations can preclude junior program administrators' attainment of a deeper understanding of the complexities of the underlife of program development and management that happens behind the scenes: for instance, discussions involving graduate students; allocation of appropriated funds; decision-making processes; curriculum development; articulating a program vision; and cooperating, collaborating, and negotiating with local administrators are tasks that must necessarily remain with the supervising faculty member or department chair. In academia, training for administrative positions is neither as deep nor as comprehensive as it ought to be in order to prepare developing administrators for their soon-to-be-realized professional realities.

We believe that this edited collection will give program directors and administrators a context for elaborating stories often told too quickly at conferences or in e-mail. In the vein of "every PTSC program administrator has a story to tell," we solicited chapters from select directors whose original stories have become significant in TSC program administration circles. For example, North Carolina State University and North Dakota State University have each created PhD programs that combine study from two departments (English and Communications). Equally impressive, Texas Tech University started the first completely online PhD program. The goal of this collection is to help current and future program directors and administrators *how to think* about the various aspects of PTSC program development and administration. This collection does not aim to be a compendium of best practices. Instead, it is a book of tales—some legendary, others ordinary—that help readers understand, in a material and direct manner, the day-to-day work of the technical communication program administrator. The work will sometimes overlap with stories told by WPAs in rhetoric and composition programs, but the stories highlight the distinctive role of the PTSC program administrator in today's academic workplace.

## SUMMARIES OF CHAPTERS

### Deborah Andrews, Stephen A. Bernhardt, Kelli Cargile Cook, Jeff Grabill, Bruce Maylath, Dan Riordan, and Stuart Selber

In "Tracing the Intellectual Trajectories of Professional/Technical/Scientific Communication: A Roundtable Perspective From Seven CPTSC Past Presidents," Deborah Andrews, Stephen Bernhardt, Kelli Cargile Cook, Jeff Grabill, Bruce Maylath, Dan Riordan, and Stuart Selber offer a historical narrative that draws together five significant story lines, focusing on threads such as pedagogy, technology, globalization, budgets, and job markets that emphasize the dynamic among social, professional, and programmatic forces, reminding us that the community both shapes and is shaped by the culture within which we live and work. Collectively, these authors examine the increasingly complex and interconnected challenges administrators face as well as the evolution of evermore sophisticated approaches to understanding and addressing those challenges.

### Nancy W. Coppola

Nancy W. Coppola narrates a story of moving from a personal story of a myopic to panoptic view of assessment practices in technical communication, which mirrors a trajectory of the field's experience with assessment. Her extensive knowledge of assessment in the field moves from the use of standardized methods to the more socially constructed models that emphasize the need for contextalized and localized practices. Her story shows, she argues, the importance of the emotional connection to those practices required of a true program assessment champion.

### K. Alex Ilyasova and Tracy Bridgeford

K. Alex Ilyasova and Tracy Bridgeford offer narratives that explore the struggles to define appropriate outcomes for technical communication programs based on our own community values rather than on values adapted from composition. The outcomes framework they construct is anchored in contemporary interpretations of *techne* and contextualized to accommodate both local institutional missions and evolving expectations of professionals in a range of work settings. The authors call first for dialog and ultimately, for action.

### James Dubinsky

James Dubinsky's narrative depicts his experience in a three-semester collaboration with a business management professor that involved visiting, participating in, and eventually reflecting on a story he sees as important to the *techne* of teaching and subsequently program building and administration. He tells his story, which is both a *techne* and a story about *techne*, as a way of knowing and sharing tales of theory in practice in order to argue for the importance of procedural knowledge of teacher-researchers.

### Meg Morgan

Meg Morgan describes the process she experienced developing from a transactional leader with personal power to a transformational leader with prosocial power. She sees the responsibility of program leaders to create communities at-large both within and outside the university in ways that empower others to be leaders who pay attention to human and social values in their relationships with others. She offers readers some ideas for enabling transformational leadership skills in ways that examine with other stakeholders the community's values, empower students with real responsibility, and create relationships around a shared enterprise.

### Ritu Raju

Ritu Raju's narrative engages the reader in a story that has been absent from many programmatic conversations—administering a technical communication program in a 2-year college. Her story of needs assessment demonstrates her college's findings from focus groups, which included hiring managers from the local Society for Technical Communication (STC) chapter and emphasizes the challenges 2-year colleges face, including relationships with industry, duration, admissions, transfer of credits to 4-year programs, and timeliness of knowledge and skills. Her story is one of reviving a technical communication program and reshaping its focus by better preparing students for technical writing positions and by raising the profile of technical communication as a sophomore-level elective.

### Laurence José

Laurence José's narrative traces the nuances of the programmatic weight of the issue of globalization in the field of technical communication through a review of all CPTSC proceedings and diversity-grant findings as reported by the grant awardees. This review provides a historical perspective of the field's approach to globalization in ways that enable us to extend the ongoing conversation and to highlight current exigencies. This diachronic story, she says, offers a constructive context for reflecting on the identity of the field and the meaning of its practices.

## M. Ann Brady and Karla Saari Kitalong

M. Ann Brady and Karla Saari Kitalong narrate the complex process of remodeling and reimagining an iconic space significant to the histories of both composition and technical communication—Michigan Tech's Center for Computer-Assisted Language Instruction. They emphasize the complexity of the linkages among institutional mission and memory, programmatic stability and change, and the transitory nature of university life for its faculty and student members. Core to Ann and Karla's narrative is the evolution of a user-centered, participatory administrative model for recognizing and addressing programmatic problems.

## Julie Dyke Ford

Julie Dyke Ford describes a unique situation at her institution. After 8 years as technical communication program director, during which time she taught a full course load and earned tenure, she now holds a joint appointment with mechanical engineering, charged with integrating communication instruction within engineering courses. Her chapter narrates the history and exigencies of this joint appointment, focusing primarily on the benefits of such initiatives.

## Pavel Zemliansky

Pavel Zemliansky describes his work as the founding director of Writing Across the Curriculum (WAC) at one of the largest universities in the country. He emphasizes how the traits and values he developed as a professional communication teacher/scholar translate into effective WAC leadership. From analyzing and responding to specific rhetorical situations, to designing appropriate communication venues, to developing and sustaining academic and administrative partnerships, to assessing the usability of these solutions, he shows how he relies on technical communication skills and sensibilities to move his program forward.

## Lisa Meloncon

Because the majority of TPC undergraduate degree programs are emphasis degrees rather than majors or minors, Lisa Meloncon argues that program administrators need a better understanding of the challenges and benefits of such programs. Using her own institution as an example and demonstrating that emphasis degrees grow out of local needs and resources, Meloncon outlines several dimensions of emphasis degrees and offers recommendations for program administrators ranging from the technical (creating a sustainable curriculum, the role of technology) to the humane (contingent faculty, being good neighbors).

## REFERENCES

Barry, D., & Elmes, M. (1997). Strategy retold: Toward a narrative view of strategic discourse. *The Academy of Management Review, 10*(2), 429–452.

Blyler, N. R. (1996). Narrative and research in professional communication. *Journal of Business and Technical Communication, 10*(3), 330–351.

Blyler, N., & Perkins, J. (1999). Guest editors' introduction: Culture and the power of narrative. *Journal of Business and Technical Communication, 13*(3), 245–248.

Brockmann, R. J. (2002). *Exploding steamboats, senate debates, and technical reports: The convergence of technology, politics, and rhetoric in the Steamboat Bill of 1838.* Amityville, NY: Baywood.

Brown, S., & Enos, T. (2002). *The writing program administrator's resource: A guide to reflective institutional practice.* Mahwah, NJ: Lawrence Erlbaum.

Dew, D., & Horning, A. (2007). *Untenured faculty as writing program administrators: Institutional practices and politics.* West Lafayette, IN: Parlor.

Enos, T., & Borrowman, S. (2008). *The promise and perils of writing program administration.* West Lafayette, IN: Parlor.

Faber, B. (2002). Professional identities: What is professional about professional communication? *Journal of Business and Technical Communication, 16*(3), 306–337.

Fearing, B., & Sparrow, W. K. (1989). *Technical writing: Theory and practice.* New York, NY: Modern Language Association.

George, D. (1999). *Kitchen cooks, plate twirlers and troubadours: Writing program administrators tell their stories.* Portsmouth, NH: Boynton/Cook.

Kreiswirth, M. (2000). Merely telling stories? Narrative and knowledge in the human sciences. *Poetics Today, 21*(2), 293–318.

L'Eplattenier, B., & Mastrangelo, L. (2004). *Historical studies of writing program administration: Individuals, communities, and the formation of a discipline.* West Lafayette, IN: Parlor.

Matthews, C., & Zimmerman, B. B. (1999). Integrating service learning and technical communication: Benefits and challenges. *Technical Communication Quarterly, 8*(4), 383–404.

McGee, S., & Handa, C. (2005). *Discord and direction: The postmodern writing program administrator.* Logan, UT: Utah State University Press.

Rose, S. K., & Weiser, I. (Eds.). (1999). *Writing program administrator as researcher: Inquiry in action and reflection.* Portsmouth, NH: Heinemann-Boynton/Cook.

Rose, S. K., & Weiser, I. (2002). *Writing program administrator as theorist: Making knowledge work.* Portsmouth, NH: Heinemann-Boynton/Cook.

Selber, S. (1997). *Computers and technical communication: Pedagogical and programmatic perspectives.* Greenwich, CT: Ablex.

Selber, S. (2004). *Multiliteracies for a digital age.* Carbondale, IL: Southern Illinois University Press.

Selber, S., & Johnson-Eilola, J. (2001). Sketching a framework for graduate education in technical communication. *Technical Communication Quarterly, 10,* 403–437.

Spinuzzi, C. (1996). Pseudotransactionality, activity theory, and professional writing instruction. *Technical Communication Quarterly, 5*(3), 295–308.

Staples, K., & Ornatowski, C. (1997). Introduction. In K. Staples & C. Ornatowski (Eds.), *Foundations for teaching technical communication: Theory, practice, and program design* (pp. xi–xxi). Greenwich, CT: Ablex.

Tebeaux, E. (1997). *The emergence of a tradition: Technical writing in the English Renaissance, 1475–1640.* Amityville, NY: Baywood.

Winsor, D. (1996). *Writing like an engineer: A rhetorical education.* Hillsdale, NJ: Lawrence Erlbaum.

CHAPTER 2

# Tracing the Intellectual Trajectories of Professional/Technical/Scientific Communication: A Roundtable Perspective from Seven CPTSC Past Presidents

***Deborah Andrews, Stephen A. Bernhardt, Kelli Cargile Cook, Jeff Grabill, Bruce Maylath, Dan Riordan, and Stuart A. Selber***

## INTRODUCTION

This chapter offers a global view of forces acting upon and being affected by professional, technical, and scientific communication (PTSC), considering the most important influences over the past 30 years or so. The chapter is modeled after *The Atlantic*'s occasional expert roundtables. For this chapter, the seven most recent past presidents (as of 2010) of the Council for Programs in Technical and Scientific Communication (CPTSC) described forces affecting the field of professional communication within the following categories:

- Pedagogy
- Technology
- Globalization
- Budgets
- Job market

We think that, as a field, we are not just affected by these forces of change, but that there is some reciprocity that, in fact, we shape these forces through the programs we construct and the teaching and research we conduct.

## PEDAGOGY EMERGING IN THE 21ST CENTURY

During the late 20th and the early 21st centuries, the acts of teaching and learning have come into sharp focus. Business leaders, government agencies, legislatures, and researchers have turned their gaze to teaching and learning. The U.S. federal government's No Child Left Behind Act, the Spellings Commission Report, and articles about the difficulty of funding local schools are examples of the ongoing discussion. As the drafting of this section began, that day's lead story in the Minneapolis *Star Tribune* was headlined, "Thousands Fail to Graduate" (Johns & Draper, 2010). Seeing the headline as it made its way down the checkout belt, the cashier said, "Looks like some teachers need to be fired." Thus, teaching and learning are in the air, and not always positively.

## THE SCHOLARSHIP OF TEACHING AND LEARNING

The Scholarship of Teaching and Learning movement is sweeping our field, just as it has others. One can trace its intellectual and professional emergence to the birth, in 2004, of the International Society for the Scholarship of Teaching and Learning (ISSOTL, 2009). Having been founded by 67 members, it grew by 2010 to thousands of members worldwide. At its annual conference in 2009, some 650 scholars participated from 500 campuses and 15 nations.

The ISSOTL's discussions raise the following questions for our field: Of the many currents that flow together to create our professional contexts, on which should we focus? What are the key trends? In this section, we focus on two trends, or currents, and the action that those trends beckon us to perform.

## THE CURRENTS OF DIFFERING TRENDS

In the 21st century, our pedagogical lives seem to rest on a sandbar amidst two parallel flows in the bend of a river: a fast outer current of calls for us to change our ways of thinking about learning and teaching, and a slow inner current, inherently conservative in nature, to view responses to those calls negatively—a current fed by the waters of our professional ecosystems and traditional disciplines.

The fast current provides the energy advancing the trajectory most prominent in pedagogy. For a number of years now it has manifested itself in a roar to create learner-centered classes. This trajectory has accelerated not only because of demands made to prepare students for 21st century realities, which are seen as significantly different from those of the 20th century, but also because of conclusions based on social-cognitive research about how people learn.

Pushed along by a global economy and a flood of technological innovations, the call to prepare students for 21st century realities comes from many upstream sources. Here, though, we focus on just one, the Association of American Colleges and Universities' (AACU) widely circulated report "College Learning for the New Global Century" (2007). In it, the AACU mentions four realities that define our 21st century lives:

- an economy in which the challenge is to innovate or be displaced,
- a democracy that is diverse, globally engaged, and dependent on citizen responsibility,
- a world of daunting complexity, and
- a period of relentless change. (p. 13)

To adapt to these realities, students must be "intellectually resilient, cross-culturally and scientifically literate, technologically adept, ethically anchored, and fully prepared for a future of continuous and cross-disciplinary learning" (p. 15).

To develop students with those qualities, many scholars have turned to the realm of the science of learning. The conclusions formed in this branch of learning in the past 40 years have indicated that major changes should occur in the way learners grow in knowledge. Bransford, Brown, and Cocking (2000), in *How People Learn*, provide one of the best summaries of this work. The three key findings on which the authors base their discussion are as follows:

- Learning is affected by preconceptions about how the world works (often simply called "prior knowledge").
- Competence in an area depends not only on factual knowledge but also on having a conceptual framework for that knowledge and methods for retrieving and applying that knowledge (often discussed in terms of "expert knowledge").
- Metacognition, or defining goals and monitoring progress in achieving them (often called "reflective practice"), causes one to take control of learning.

From these findings come calls for action. Bransford et al. (2000) called for teachers to create environments in which learners immerse themselves in activities that follow from those conclusions. That environment is learner-centered because its teachers focus on the attitudes, knowledge, and skills that students bring to the classroom. It is knowledge-centered because students manipulate data and concepts such that they organize knowledge around concepts and learn to transfer learning to other situations. It is assessment-centered because teachers provide regular formative (or "in-process") assessment. It is community-centered because it fosters a camaraderie that encourages problem solving and question asking (pp. 8, 23–25). The goal of such immersion? That learners create "usable knowledge" that can be transferred to other situations.

The AACU (2007) urged teachers to facilitate "intentional learners." These learners focus on "big questions" in all disciplines. They practice intellectual and practical skills in the "context of progressively more challenging problems, projects, and standards for performance" (p. 3). They are actively involved in diverse communities and real-world challenges. They are "integrative," that is, able to apply knowledge, skills, and responsibilities to new settings and complex problems (p. 3).

With these calls to action has come a wealth of books and articles on how to develop this learner-centered world. Just a few include Tagg (2003) *The Learning Paradigm College*; Fink (2003) *Creating Significant Learning Experiences*; Wood (2009) *Lecture-Free Teaching*; Huba and Freed (2000) *Learner-Centered Assessment on College Campuses*; Walvoord and Anderson (2010) *Effective Grading*; Wieman, Perkins, and Gilbert (2010) "Why Not Try a Scientific Approach to Science Education?"; and Ambrose, Bridges, DiePietro, Lovett, and Norman (2010) *How Learning Works: Seven Research-Based Principles for Smart Teaching.*

These authors asked that teachers—but also colleges, programs, and departments—change their approach in fundamental ways. The classroom must become learner-centered and, as Jim Dubinsky (2004) articulated in his excellent introduction to *Teaching Technical Communication*, teachers must become "user-centered, reflective practitioners" (p. 4). This type of classroom functions differently from lecture-based classrooms. Learner-centered classrooms provide problem-based, inquiry-based, active learning activities. Learners work with their preexisting experiences, manipulate data and concepts, and reflect on how and why learning occurred. Moreover, in the programs that follow this trend, opportunities exist for students to participate in what the AACU called the effective educational practices: first-year seminars and experiences, common intellectual experiences, learning communities, writing-intensive courses, collaborative assignments and projects, undergraduate research, diversity and global learning, service learning, internships, and capstone projects (2007, pp. 53–53, 61–62).

However, the university tends to float in the slower, conservative current. Although many individual instructors are convinced of the value of the learner-centered approach, many departments and their promotion committees are not. Commenting on teaching in science, Carl Wieman (2007), a Nobel-winning physicist, said,

> Many internal institutional programs have achieved little overall change in Science Technology Engineering Mathematics (STEM) teaching at the large research universities. This remains a major problem for improving

> science education at any level, as these universities largely set the norms for how to teach science and what it means to learn science. (p. 8)

John Tagg (2003) maintained that many colleges and universities operate under an "instructional" paradigm that focuses on semesters, lectures, coverage, final grades, student GPA, and faculty excellence usually defined by research output. Tagg maintained that colleges and universities must transform themselves by creating a "learning" paradigm. In this paradigm, students focus on pursuing their goals, give frequent performances, receive frequent feedback, belong to stable communities of practice, and see learning as having a long horizon rather than the arbitrary semester of 14–16 weeks (p. 124).

Converting to the learning paradigm requires a significant change not only in individual teachers but also in departments, colleges, and organizations as a whole. An individual, especially tenured professors, can make the changes needed to encourage a learning paradigm mentality in their class. But often, when an untenured assistant professor devises a community-based class of active learning, aimed to develop students who have usable knowledge and who can interact effectively with the complexities of the 21st century, problems arise: Will an assistant professor who teaches in that manner be retained? Will teaching in that manner be accepted by the culture of the department? And what if it is not? Weiman (2007) attacked the issue by creating a "Science Education Initiative" at the University of British Columbia. Departments applied competitively to receive a substantial grant that enabled (and required) the entire department to undertake the transformation of every course to a learner-centered approach. His reasoning:

> The department is the unit at a research university that decides what is taught and how it is taught in that discipline; thus, any sustained attempts to change teaching practices must focus on the culture of the department. To change that culture, one must affect most undergraduate courses and involve most faculty members. (p. 8)

He reports a great deal of progress and significant improvements in teaching practices. However, even with these successes, Wieman did not address such issues as departmental and university policies and standards for promotion and tenure. Unless those issues are also addressed, the transformation of the learning culture will be inhibited.

These two currents—learner-centered teaching and institutional cultures—feed the waters on which we float as the 21st century enters its second decade. Our success at navigating over to learner-centered strategies, both in our individual teaching and in our departmental and university policies, will determine whether our pedagogical vessel will propel us through the rapids that await or leave us floating in the same old place. The pedagogy of Professional, Technical, and Scientific Communication (PTSC) has had a long history of active learning but less, perhaps, of learner-centered classrooms. As we join the work already begun, we must become what Dubinsky (2004) called "user-centered, reflective practitioners" (p. 4). We need to think through our concepts of the ethos both of ourselves and of students in the classroom, develop the courage and skill to put into action and into policy those ideas, and acquire the know-how to create the 21st century student that our colleges and our nation desire.

## PROFESSIONAL COMMUNICATION AND TECHNOLOGY

The field of PTSC has an especially rich history of debate about the curricular integration of computers. For three decades, the field has been preoccupied by a series of questions that continue to challenge us in interesting and meaningful ways:

- Should we teach software skills, and if so, how?
- How should we design our workspaces, particularly computer labs and classrooms?
- How much of our teaching can be conducted online?
- How can computers aid the tasks of program administration?

What makes these questions significant is that they involve not only particular instructional or program practices but also conceptions of the profession itself. In other words, questions of technology are also always questions of identity: We define ourselves as a field, at least in part, through our stances toward, uses of, and relationships with technology. Through the mediation of networked computers, we represent and share values and practices, build relationships, and increasingly, structure the shape and flow of programmatic work. Before networked computers, these tasks were not so immediately visible and not so thoroughly collaborative, especially on national and international scales.

There are, of course, multiple valid answers to questions about the curricular integration of technology. For both social and technical reasons, an approach from one program might not work very well elsewhere, or it might unfold and function in very different ways. This reality, which illuminates inevitable tensions between global and local cultures, can make it difficult for the field to orchestrate a larger dialogue about technology development and use or for one program administrator to interpret and evaluate advice from another. We continue to see real value in sharing individual program experiences, for such experiences do not have to be generalizable to be instructive and useful. But we also want to encourage colleagues across the field to articulate broad habits of mind that can guide discussions and decisions—at all levels, from local to global—in directions that contribute positively to our professional identity and status and that position the field to contribute to institutional initiatives that affect us, but do not always include us, as participant-designers. These habits of mind involve specific definitions of technology and specific understandings of its relationship to institutions.

## DEFINING TECHNOLOGY

What is technology? More to the point, what is a productive characterization of technology for PTSC? Although some have included technology in their definitions of PTSC (e.g., Dobrin, 1983), more often than not the field concentrates on other aspects of the rhetorical situation, relying on various taken-for-granted assumptions about what technology is (and does) in culture, academic and nonacademic. One common move in our scholarship is to equate "technical" practices with "specialized" practices, to expand the domain of PTSC to include a much wider range of disciplines and activities. However, we should also be clear and comprehensive about our perspectives on technology development and use. After all, these perspectives shape, in a central fashion, how we imagine the nature of communication.

In our view, technology is an "articulation" of sociotechnical forces that functions to rationalize symbolic action in PTSC. (For more on the concept of articulation and its relevance to the field, see Johnson-Eilola, 2005.) By "rationalize," we simply mean structuring systems that organize work and work activities. Several points here warrant elaboration. Although a technology can certainly be a physical artifact, it can also be a process, procedure, or system. Scheduling systems are indeed technologies, but so are the protocols for scheduling classes in a university setting. In addition, the forces that constitute a technology can derive not only from technical contexts but also from social, political, and economic contexts. Consider the technology of Web design in a PTSC program: It involves software environments, style guides, institutional policies, legal considerations, collaborative methods, support systems, budgets, and more. Such forces, as articulations, coalesce in particular ways from project to project, exhibiting varying degrees of intensity across time and space. Style guides matter more to some projects than to others, for example, and can become important and influential at different moments across projects. Finally, because there are always multiple ways to structure action, our definition acknowledges the nonneutrality of technology. It is no news to report that the technologies of PTSC embody human interests and motivations. When users open a new computer, connect to an online service, or purchase a piece of software, the designers of those environments have already made a wide of range of decisions for such service or software. These decisions influence how users read and write, how they collaborate, how they share files with one another, and which program features are obvious and which ones are buried deep in the system. Acknowledging design biases, however, does not have to be seen as critical in a negative sense. We see design biases as inevitable, even necessary, and as a feature of technology to be leveraged and explored.

So, in our view, technology is a process, as well as a product, and is social all the way down to its fundamental level. Moreover, it is dynamic in nature, constantly evolving in concert with a whole host of elements and forces in settings of use and design. Studying a piece of PTSC requires us, then, to suspend ongoing interactions, to take a snapshot of one particular articulation of technology, and to use it for assessment purposes. In addition, if that assessment suggests revisions, we need to consider more than the formal features of the artifact or system, because those features become meaningful only in relation to the conditions of the task at hand. Thus, we perceive a robust vision for the field: technical communicators as architects of sociotechnical contexts.

## DEVELOPING PRODUCTIVE HABITS OF MIND

Our professional conversations often concentrate on either teaching with technology or designing degree programs to support technology initiatives. Although these conversations address issues everyone faces, their scope of usefulness tends to get narrowed by the conditional and contingent nature of technology in institutional settings. In response to this situation, we offer five "habits of mind" as a conceptual adjunct to the impulse to focus on specific pedagogical and programmatic practices. As a whole, these habits function as a heuristic for examining relationships between technologies and the institutional sites that house PTSC programs. The first two habits apply to analysis activities; the last three, to production activities.

### Mapping Institutional Practices

How many of us are students of our institutions? How well do we understand the dynamics of the places within which we work, particularly those associated with technological design and decision-making? We want to encourage teachers and administrators to study how their institutions plan, develop, support, and manage technology. In our experiences, having a big-picture view is crucial to creating strong, sustainable frameworks that can support program activity. For example, in a study of the computing culture at Pennsylvania State University, Selber (2009) discovered that his administrative approaches did not always align with valued patterns of resource allocation or with instructional-design models for distance education. He responded, in part, by rearticulating his approaches, but he also used the results to make the case for new and different patterns and models at the institutional level. In a similar study at Michigan State University, DeVoss, Cushman, and Grabill (2005) concluded that "writing programs will never adequately come to terms with how to understand and teach new-media composing unless we can come to a productive and activist understanding of infrastructure" (p. 22). This awareness, we would argue, may be best engendered through organized efforts at mapping institutional practices. The studies by Selber (2009) and DeVoss et al. (2005) discussed spatial techniques and heuristics that can guide this sort of intellectual work.

## Analyzing Literacy Claims in Formal Features of Technology

With this habit, we advocate a very particular way of conceptualizing the constructed character of technology. We find it profitable to think about the formal features of artifacts or systems as a series of claims about the nature of literacy and work (see Carroll & Rosson, 1992). For instance, the navigation and annotation features of e-book devices are informed by formalized assertions about the reading practices of users. Likewise, the collaborative features of course-management systems are informed by formalized assertions about the pedagogical practices of teachers. Stated or otherwise, such assertions are an inescapable component in the process of creating design representations, bringing order and coherence to an artifact or system, and structuring its affordances in concrete ways. In their study of the Sony Reader, Faris, Lundin, and Selber (2010) noticed a direct relationship between the claims made by a technology and its acceptance rate by users. In short, users reacted favorably to e-book features whose claims more or less aligned with their reading practices and beliefs about reading. This finding, although hardly surprising or unexpected, is powerful nonetheless, reminding us that institutional approaches to technology adoption, development, and use should aim to make explicit the literacy claims of both designers and users. Understanding these claims is crucial to both functional and critical program tasks, and in institutional terms provides teachers and administrators with a unique perspective to contribute to larger conversations.

## Employing Recursive Design Practices

As a field, we tend to respond to questions about the curricular integration of technology with rather linear approaches. We work backwards from the job market, for instance, identifying skillsets that students need as workplace practitioners and then designing institutional infrastructures for developing those skills. Or, in the context of creating a technical communication course, we first define our instructional goals and then match those goals with educational technologies that can help us achieve them. Although many members of our field are persuaded by rhetorical perspectives that contextualize all aspects of discourse production and use, linear approaches do not account for the ways in which institutions and technologies shape each other: The affordances of computers can encourage teachers to reimagine their pedagogies, for example, and the infrastructures of academic programs can influence the direction of skill development.

Rhetorical approaches, as they have been traditionally mapped out, do not tend to address this dialectical relationship. We encourage teachers and administrators to employ recursive design practices and to be sensitive to the interanimations between technologies and institutions.

## Equating Writing Tasks and Technology Tasks

How much time do instructors devote to teaching technology skills? How do instructors keep technical communication courses from becoming courses in software applications? These are routine questions in the field, and important questions, but we would challenge strict distinctions between the tasks of literacy and the tasks of technology. On the one hand, we are sympathetic to concerns about vocational training: We would not endorse substituting a course in coding for a course in writing or communication, nor would we endorse a curriculum that does not cover the full spectrum of issues students need to learn about. On the other hand, tasks like designing style sheets, creating e-mail filters, and managing content databases are part and parcel of the discursive work of PTSC. If students need to learn to control the level and shape of language, they also need to learn to control the environments that organize and render their messages for users. Contemporary communication contexts have intensified the relationship between literacy and technology. Literally put, one does not work without the other. From a pedagogical standpoint, the problem is that institutional training programs all too often separate the why from the how, stripping context, rationale, and principles of design from discussions of technical procedures. Students will be shown how to create a style sheet, for instance, but learn precious little about typography, hierarchy, or grid systems for the page or screen. In all likelihood, they will be advised to default to the themes provided by the software program. Our assignment, then, is to recontextualize technology instruction with technical communication instruction, to situate the teaching of functional tasks in social terms. As a field, we must work to create and share approaches to this worthwhile assignment.

## Valuing Micropolitical Actions

This last habit has to do with the nature of agency and change in institutional settings. As discussed in the previous section on pedagogy, the bureaucracy and glacial pace of academia, the primary work site for programs in PTSC, stand in vivid contrast to the dynamic enterprise and rapid pace of technology development use. For both better and worse, institutions create structures that stabilize technology and enable it to function in a systematic and coherent manner. These structures, however, can feel like challenges or barriers to change, especially for teachers and administrators working in unconventional institutional directions that, of course, evolve and shift with time. (Wikis were once considered to be unconventional institutional technologies, as were channels for Internet relay chat.) Teachers and administrators should obviously work to create lasting campus change, advocating for sociotechnical arrangements congruent with the directions of the discipline. But we also want them to value micropolitical actions, actions that take place in the interstices of formalized institutional structures. Although often contingent and temporary, micropolitical actions can have a real, direct, and positive effect on the people in our programs. For a striking example, consider the impact of Creative Commons licensing on campus discussions of copyright law and on pedagogical practices for multimedia. Other examples from our experiences include using the features of quizzing software to collect class feedback, replacing institutional metaphors for technology with metaphors that emphasize human as well as technical aspects, and using the surveillance features of course management systems in ways that help students in distance classes feel more connected to their teachers and peers. By physical or discursive means, micropolitical actions can help to align the contexts of institutional technologies with the values and practices of PTSC.

These habits of mind are important for teachers and administrators to develop. Although far from exhaustive, they are fundamental to thoughtful approaches to the continuing trajectory of technology's influence on professional programs.

## GLOBALIZATION

Though the global economy is hardly news, aspects of it are, particularly its transformation of the work environment and communication technologies. This transformation prompts new questions and new approaches for technical communicators and those who teach and conduct research in the field.

### Eras of Transformation and Questions Raised by the Current Era

Though not without flaws and critics, Friedman (2005) provided a useful and oft-cited analysis of three phases of transformations in the nature of global work. In the first, Globalization 1.0, following Columbus's opening of trade between the eastern and western hemispheres, *countries* sought out a global economic stage. Religious or imperialistic (or both) motivations prompted exploration; natural (including human) energies powered it. During Globalization 2.0, from roughly 1800 to 2000, *multinational companies* drove global trade. Fossil-fueled transportation technologies, coupled with communication technologies (telegraph, telephone, television), and a range of financial instruments and institutions, catalyzed extensive international economic growth. Then around 2000, Globalization 3.0 began. In this era, *individuals*, rather than nations or companies, are the main agents of change. Lean global organizations are marked by flattened hierarchies, less supervision, more focus on individual entrepreneurship, and virtual teamwork among geographically distributed work groups. Their work often involves products and systems designed in one location and manufactured elsewhere.

Although this picture does not represent all work, its broad lines suggest that academic programs, pointedly in the United States, require reorientation to a new trajectory by assuming what European programs have assumed all along; namely, that all professional communication is international. Everyone, not just those in an international unit of a company or a government's foreign office, must operate in the 3.0 version of the globalized work environment. Our curricula and textbooks must interweave international concerns throughout, not in separate courses or units or chapters (Maylath & Thrush, 2000). When they are graduated from academic programs and step onto a (relatively flattened) international playing field, practitioners of PTSC, out mostly on their own, must find their voice and set their ethical compass among people not like them, in districts with shifting rules and values. What expertise does this global setting demand, and how can we prepare students to thrive in such a setting?

Underlying Globalization 3.0 as well are rapid changes in communication technology. As Holmes (2008) so thoroughly documented, we have moved from a time of amateur scientists through a period of increasing professionalization from the late 19th century through the 20th century. New media, on laptops and powerful hand-held devices, have ushered us into an age of "citizen scientists," of "crowd sourcing," in which science's march to the answer can involve contributions from thousands of individuals worldwide. In a social networking environment where everyone can be an equal contributor—for example, Wikipedia, where expertise is distributed and authority increasingly questioned—what is the role of the professional writer/editor?

Finally, in this obviously selective list of transformations in the global workplace, questions arise regarding the renewed emphasis worldwide on sustainability and green living. Increasingly we recognize that we live together on a single planet, where ash from an erupting volcano in Iceland can bring transport worldwide to its knees and where any individual act, say, on an offshore oil platform, can have unintended consequences both near and far away. How can PTSC practitioners foster appropriately sustainable thinking and behavior? How do we prepare them to aid in the design of products to be environmentally friendly as well as self-explanatory (no need for translation, no need for manuals)?

The questions concluding each preceding paragraph prompt no simple answers. Reevaluating the components of expertise is an exercise in which the CPTSC has been engaged for years in meetings in the United States and abroad. Certainly one component is an enhanced understanding of the diversity of peoples, cultures, and languages. Another is facility with communication technologies, understanding what they can and cannot do. In a setting of crowdsourcing and citizen science, the technical communicator plays a role akin to community organizer. The technical communicator adds value by framing the right questions and finding the right problems, sorting through the various contributions, and measuring the worth of each while focusing them toward the problem at hand.

## Study Abroad[1]

To develop students' intercultural competencies and promote their success in the global economy, more and more programs worldwide have encouraged their students to study abroad. In the decade from the 1994–1995 academic year to 2004–2005, the Institute of International Education (n.d.) reported a 144% increase in enrollment in study abroad by U.S. students (in raw numbers, from 84,403 participants to 205,983). Once limited in the United States mainly to students of foreign languages, such periods of study abroad now attract students across a wide spectrum of majors, including students of PTSC.

To accommodate rising demand, U.S. colleges and universities are developing programs for delivery abroad that are innovative in both their time frame and their content. Students in business and technical fields often find that their tight curricula and work obligations make studying abroad an obstacle in their progress to the degree. In addition, U.S. administrators are adding fees and other financial requirements as they see revenues drop when universities abroad or programs run by other universities draw off students through lower tuition and housing costs. Some universities require students to pay full at-home tuition, even if the host school abroad costs less, and cap the financial aid available for study abroad (Bernstein, 2006). To meet the needs of both students and administrators, international program offices at many schools encourage their own faculty to develop short courses in a variety of subjects, including technical and business ones that can be taught abroad. In fact, according to the Institute for International Education (n.d.) Open Doors 2006 report, the majority of students (56%) chose to participate in courses lasting less than one semester (the fastest growing option); 38% studied for a semester; and 6% for a full year.

Shorter study-abroad courses offer advantages for students as well as faculty. Students who would otherwise not be able to afford the direct costs (program fees, food, and lodging) and indirect costs (lost wages from work) of a full-year or full-semester program may be able to participate in a shorter program. So can faculty who might otherwise not be able to take the time away from their regular-semester teaching responsibilities. Instead of sending students abroad to study at other institutions, these programs provide ways for the faculty at U.S. institutions to accompany their students abroad, enabling greater university control over both revenues and curricula.

[1] Portions of this discussion of study abroad were originally published in *Business Communication Quarterly, 72*(1) (March 2009). We are grateful to the Association for Business Communication for permission to reprint that text.

## International Collaborative-Learning Projects

In addition to the growth of study abroad as an integral component of PTSC curricula, academic programs have begun to experiment with projects that bring students from different countries, cultures, and languages into collaboration with each other without physical travel. Such experiments respond to globalization through the one key factor permitting the quick expansion of globalization in the first place: instant communication technologies. Using everything from e-mail to websites, wikis, iChat, and various other real-time links with streaming video, universities hundreds of kilometers apart began pairing their classes in conversation to complete joint assignments. Many of these efforts, some of which have been operating for over a decade, are detailed by their creators in Starke-Meyerring and Wilson's (2008) anthology, *Designing Globally Networked Learning Environments.* Some of the examples referenced are in neighboring countries such as collaborative-learning projects involving students between the University of Delaware and Canada's McGill University or between The Netherlands' Radboud Universiteit Nijmegen and Belgium's Lessius Hogeschool. Most, however, cross seas or oceans, such as between Fairfield University in Connecticut and Nicaragua's Universidad Centroamericana or between the University of Washington and The Netherlands' Universiteit Twente or between the Georgia Institute of Technology and Russia's European University and Volgograd State University and Sweden's Blekinge Tekniska Högskola, or in one of the most elaborate networks, dubbed the Trans-Atlantic Project, the University of Wisconsin-Stout and North Dakota State University with Austria's Karl-Franzens-Universität Graz, Belgium's Hogeschool Gent (University College Ghent), Denmark's Århus Universitet, Finland's Vaasan Yliopisto (Vaasa University), France's Université Paris-Denis Diderot, and Italy's Università degli Studi di Trieste and Università degli Studi di Padua.

## Merging Technical Communication and Translation

Perhaps one of the most notable developments related to globalization is what Gnecchi, Maylath, Mousten, Scarpa, and Vandepitte (2011) have identified as "field convergence": the merging of the historically distinct roles of technical communicator and translator. Their surveys reveal that translators in particular are seeking cross-training so that they can compose technical documents from scratch and thus capture the work (and payment) that used to go to technical communicators before translators received the documents. The trend is especially noticeable in Europe, but is detectable in North America as well. It is driven in part by the acceleration of the documentation process such that, with the help of content management systems, technical documents are composed in 30 or more languages simultaneously, thus obliterating translators' earlier distinctions between source text in source language and target text in target language. The implication for academic programs seems clear: There needs to be increasing cooperation between technical communication and translation degree programs, including the possibility of joint degrees. The blooms of such cooperation can already be seen budding in such places as the University of Washington, whose Human Centered Design and Engineering Department (formerly titled Technical Communication) includes a certificate program, an undergraduate minor, and a master's degree in technical Japanese; and the University of Wisconsin-Milwaukee, whose English Department and Foreign Languages and Literature Department have worked together to offer a graduate certificate and a master's degree in translation (French to English, German to English, Spanish to English, or English to Spanish).

### Professional Organizations

Responding to globalization, professional organizations have become more international. The Society for Technical Communication (STC) added chapters beyond its traditional base in the United States and Canada. Current chapters circumnavigate the globe from Belgium to Israel to India to Japan. Moreover, the STC joined with counterparts in other countries, such as the United Kingdom's Institute of Scientific and Technical Communicators (ISTC) and Germany and Austria's tekom, to form an umbrella organization called INTECOM. From 1975 to 2003, INTECOM sponsored a series of international conferences called FORUM. In a separate move, many of INTECOM's constituent organizations based in Europe formed TCeurope. Since 2001, TCeurope has sponsored yearly colloquia hosted in one of the member countries. In turn, the International Professional Communication Conference (IPCC), sponsored by the Professional Communication Society of the Institute of Electrical and Electronics Engineers, first met outside the United States and Canada in 1990, when it convened in England. It did not do so again, however, until 2005, when it met in Ireland. Thereafter, its meetings have taken place with some regularity in different nations and continents, such as The Netherlands in 2010 and tentatively Japan in 2012. The North America-based Council for Programs in Technical and Scientific Communication and the Association for Teachers of Technical Writing took advantage of Intecom's FORUM and IEEE-PCS's IPCC by sponsoring their own joint roundtable series for educators. Beginning with the FORUM in London in 2000, the roundtables have varied from all-day to half-day to evening events whenever FORUM or the IPCC has been held outside the United States.

## HIGHER EDUCATION FINANCING AND THE CASE OF PROFESSIONAL, TECHNICAL, AND SCIENTIFIC COMMUNICATION

It is not news to anyone that higher education funding is in a very challenging period. Many of our programs are in public universities—and not the flagship universities but the land grant school or the normal college that has become a public university. Such schools traditionally rely more on state funding, have less ability to attract major research funding, and face a more difficult time during recessions. Faculty members at institutions better able to attract funding are under increasing pressure to do so. Whereas public universities were once funded primarily through allocations from the state, public allocations represent a declining percentage of the overall budget (with North Dakota seemingly the exception that proves the rule)—not only in the United States but also the UK, Canada, and most of Europe. In some situations, state allocations are declining in real dollars, pounds, and euros, while in others, the modest increases from the state do not keep pace with the growing overall budget, so the public proportion declines.

The schools in this predicament tend to face lower endowments and fewer prospects for attracting the kinds of endowment income that support major universities through difficult times. But even at institutions with historically higher levels of funding from more diverse sources, the new economics of higher education present significant challenges. To echo the theme of Richard Florida's (2010) book *The Great Reset*, it is necessary to recognize that past funding models are giving way to new ways of thinking about higher education in general and its funding in particular. In the new economy of higher education, writing programs and their departments cannot "nickel and dime" their way to solvency, let alone excellence. Viewing the economic problems that face us in terms of scarcity and focusing on budget cuts as the only solution are sure ways to do lasting damage to a writing program. If these issues of economics and finance are being "reset," then what might such a reset look like for higher education and our programs? More to the point, what *should* they look like, and how can we participate in funding debates? It is worth considering what schools can do to stay afloat (or thrive) during difficult times. It is worthwhile for those of us who run PTSC programs to understand the current financial picture and to think about how best to respond in the interests of maintaining strong programs.

## Who Subvenes Whom?

Arguments are surfacing in various quarters that suggest, perhaps counter-intuitively, that the cheap and efficient humanities and social sciences actually provide the budget support that funds the expensive science and engineering enterprises (Newfield, 2009). It is widely thought that universities make their real money on research contracts and associated overhead. But Robert Watson (2010) of UCLA argued the contrary:

> According to spreadsheet calculations done at my request by Reem Hanna-Harwell, assistant dean of the humanities at the University of California at Los Angeles, based on the latest annual student-credit hours, fee levels, and total general-fund expenditures, the humanities there generate over $59 million in student fees, while spending only $53.5 million (unlike the

> physical sciences that came up several million dollars short in that category). The entire teaching staff of Writing Programs, which is absolutely essential to UCLA's educational mission, has been sent firing notices, even though the spreadsheet shows that program generating $4.3 million dollars in fee revenue, at a cost of only $2.4 million.

However, even if this observation reflects general financial facts across institutions and systems, any arguments on the basis of this situation should be used cautiously. The fact that teaching loads in the humanities and social sciences might support activities in other parts of campus might not provide as persuasive an argument as we might hope. The response from administration, the larger university community, and the public might very well be, "Who cares?" That is, in addition to keeping an eye on costs, we also must keep an eye on value. If value on our campuses is understood to be generated elsewhere, the fact that we are inexpensive and productive in terms of generating credit-hour-based revenue might not matter much. Furthermore, we should pay attention to how "we" are generating this revenue. "We," at least at public institutions, do so largely by commanding a meaningful share of tuition dollars and general funds from the state. It is precisely these revenue streams that are under the most pressure. How much more can students bear in terms of tuition increases? We must also consider that "we" don't exactly create student demand, at least not all of it. Students come to our institutions for many reasons, and the structure of the curriculum funnels many students into our departments and programs. The funnel can easily be redirected.

## New Budget Models

The transition on many campuses is toward responsibility-based budgeting (RBB), also known as Responsibility-Centered Management (RCM), activity-centered budgeting, or some derivative name. Under these budget models, pioneered by schools such as the University of Michigan (Courant & Knepp, 2008) and the University of Pennsylvania,[2] budget dollars are expected to flow toward centers of activity. Presumably, the whole operation becomes more entrepreneurial with tuition dollars following students to colleges and then, one hopes, to departments. For the typical PTSC program with heavy service commitments, this scheme results in a time of reckoning in terms of who does the teaching and where the money flows.

These new responsibility-based budgets give us unprecedented opportunities to address the deep structural problems of labor exploitation in the teaching of writing in universities. RBB will draw light to the questions of whether teaching writing is a subsidized activity, requiring budget subventions, or a profitable activity; and if so, under what faculty configuration? Can we afford

[2] See RCM Model at http://www.budget.upenn.edu/rcm/index.shtml

to teach writing to classes of 20 students? What about 25? 30? What would be the workload and commensurate salary of a continuing faculty member generating sufficient revenue? RBB gives us metrics to build our enterprise and justify our activities.

An advantage for our programs is that they attract and serve students from all disciplines. Most RBB models allocate funds from tuition based on the home colleges (or faculties, as they are known in Canada and Europe) of students: If a technical writing course taught by a member of the English faculty enrolls students mostly from the school of engineering or agriculture, English faculty enjoy a higher rate of return than would be the case for teaching students from their home college or faculty. In general, RBB models encourage departments and colleges to increase enrollments. Writing is a field of elastic demand: Its faculty can seek new audiences of students interested in science writing, medical writing, or business communication.

Under old budget models, there was no particular incentive to seek more students—it was hard enough to staff courses that were already required. Under old models, department chairs would see that student demand exceeded available seats and petition the dean for additional funds. The dean, in turn, would press the provost for special one-time funding to add sections (and this special "one-time" funding tended to happen every year). Under new models, the funds flow to the departments that do the teaching; at least, that's the operating theory. The flow of funds can be diverted in many ways, from chancellor/president to provost to dean to special projects or institutes. All administrators will echo the same caution, namely, that university funding cannot automatically be distributed by formula but has to be allocated through wise, prudential decision-making to protect high-value, high-cost activities. Thus, new budget models establish a set of rules and also attempt to build in flexibility to protect what is deemed most important. Whatever model is used locally, it behooves department chairs and directors of PTSC programs to understand the budget formula and to seek to align department activities with the flow of funding. (For in-depth examples, see Maylath, Grabill, & Gurak, 2010.)

## Funds From Research

PTSC faculty members can pursue external funding, either on their own or as part of a team. External grants provide funding for research activities, portions of salaries, released time from teaching, equipment and software, supplies, printing, and travel. Importantly, research grants also generate overhead, some portion of which is often returned to departments, programs, or individual faculty members. Overhead rates on federal research grants in the United States are negotiated by each institution, with rates as high as 60% of the total project cost not out of the ordinary. Training and development grants tend to have lower overhead rates on the order of 8%–12%.

Although we may be able to pursue competitive funding, the money available to us as principal investigators is often focused on a few programs that present challenges. For instance, in the United States, federal government funding for the National Endowment for the Humanities (NEH) has hovered around the $140 million level in recent years.[3] In fiscal year 2006, just over 13% of NEH funding was dedicated to "research," though the percentage available to grant programs of relevance to the humanities is much higher. In contrast, the NEH's counterpart in the sciences, the National Science Foundation (NSF) has an annual budget of about $6.9 billion (FY 2010) and is the funding source for approximately 20% of all federally supported basic research.[4] Clearly, and not surprisingly, the NSF has significantly more research funding available. Given these numbers, PTSC researchers in the United States would be wise to compete for NSF funds. However, they should enter such competition only if their disciplinary identities and values permit them to do so. That is, federal research funding in the United States is a disciplinary game, and those of us who play it have to be able to present ourselves as members of a fundable discipline if we are to succeed.

At the same time, the faculty associated with writing or rhetoric have a related identity problem at the NEH, whose administrators are keenly aware of its budget's relatively small size. The NEH has made clear that it seldom views PTSC researchers' (or general compositionists') work as part of the humanities but more often as part of the educational, social scientific, or technical professions. Each characterization creates problems. The NEH presents PTSC researchers with increasing competition for a tiny pool of research funds; the NSF presents an already highly competitive environment in which we find ourselves confronting significant identity issues. We can compete and win, but the costs of proposing are high, both in terms of the time needed to assemble competitive proposals and the frequent number of failures often required before succeeding.

Not surprisingly, research is valued above all else at research institutions. Thus, PTSC programs and departments manifesting a clear and recognizable value added by research stand the best chance to succeed. Writing-service programs are in serious trouble, even if the university still needs the service. How well are PTSC programs articulated with their universities' mission and strategic goals? At some institutions, there is virtually no connection; at others, the program is strongly articulated with the goals. It is time for programs to work toward goals of articulation—multiple and valuable connections to resource-generating activities.

[3] See American Academies of Arts and Sciences National Humanities Indicator Report at http://www.humanitiesindicators.org/content/hrcoIVA.aspx#topIV1

[4] See http://www.nsf.gov/about/

## Funds From Entrepreneurial Activities

Where else should PTSC programs look to generate their own revenues? What can we generate in terms of professional education or certification programs or short courses in the workplace? Can usability labs, workforce training centers, publishing centers, or online education become a more prominent way to generate funds for programs? All of these have been and are being developed out of core professional writing programs. For example, both Texas Tech University and Clemson University have usability labs run from within their PTSC programs.

Budget models like RBB encourage programs to be entrepreneurial. At the University of Delaware (UD), any funds from professional education return to the department. Thus, the department has important incentives to undertake new, tuition-generating models of graduate education. UD is busy creating new training and graduate education offerings for health care, military, and business professionals. The persons leading these initiatives would like oral and written communication to be a part of all professional programs. The expectation is that the training is likely to be offered in a model different from three 3-credit courses: perhaps as a series of workshops or embedded in other courses or as some sort of certification program. During recent hiring discussions at UD, the dean to which the professional writing program reports made it clear that new hires would come with the expectation that new faculty members will have a role in graduate professional education, that is, that they engage in tuition- or fee-generating activities.

How well do PTSC programs fare in institutions that are not land grants, polytechnics, or others of like mission? If those institutions are in trouble (and they may not be in anymore trouble than others), should programmatic missions be rethought?

## The Job Market

In light of the trajectories that we have delineated previously, it seems increasingly obvious that the traditional market for PTSC graduates, particularly those who become practitioners with associate, bachelor's, and master's degrees, has shifted radically and perhaps even disappeared. The technical communicator of the 1990s dot-com telecommunications boom is gone, just as the technical communicator of the typewriter era disappeared earlier still. The older faculty members among us recall teaching students how to space all-caps headings on a typewriter. Most of us recall training students to develop printed computer manuals: quickstart guides, set-up instructions, tutorials, and reference manuals. Students who had mastered these genres could then go off to IBM or Microsoft, Apple or NEC. In those days, we debated whether we needed to teach RoboHelp and require a course in programming. (IBM used to insist on at least one course in Assembler.) The activities and functions may persist, but we need to address the possibility that the job types and career trajectories that many programs target have moved, changed shape, or evaporated.

Where are we now? Documentation has moved online, into content management systems and document databases. Outsourcing has moved many jobs in technical writing, medical writing, and technical support to developing countries. The trajectory of the job market for graduates might thus suggest that programs and their graduates must continually shift to keep abreast of market trends. Yet how we understand the trajectory of the job market depends on what we mean by "disappear" or "evaporate." PTSC graduates are still finding work, including as technical writers, but often in jobs that did not exist 5 or 10 years ago and in surprising domains. Frequently they are hired to be professionals because they can communicate, and not precisely to be professional communicators. Fewer are engaged in documentation/user assistance or similar tasks as they existed 10 or even 5 years ago.

Because the market has expanded and changed, rather than retracted, the prospects for graduates remain excellent in spite of the 2007–2010 Great Recession. The most current edition of the Bureau of Labor Statistics' *Occupational Outlook Handbook* (2010–2011) includes "technical writer," also called "technical communicator," as a distinct occupation for the first time. Prior to this edition, the occupation was included under the more general occupation of "writers and editors." Employment of technical writers, according to the current edition, is expected to increase by 18% between 2008 and 2018, an increase faster than the average for all occupations.[5] Employment of technical writers, the article also notes, has expanded beyond traditional positions in technological fields and now includes jobs in the sciences, engineering, nonprofit, and public policy domains. Regardless of the domain in which they are employed, technical writing graduates are highly valued practitioners because of their communication and technological skills as well as their expertise with multimedia and Web production.

Much of the credit for the newly designated occupation of "technical writer" belongs to the Society for Technical Communication (STC), which launched a multipronged effort to build technical communication as a profession in 2007. With Professional, Technical, and Scientific Writing now designated with a standard occupational code, the movement toward professionalizing the field continues to gain momentum. Given these changes, our programs will need to begin to consider how they will participate in, shape, and eventually adapt to this professionalization movement in the field. If successful, two additional prongs in STC's efforts to professionalize the field may likely affect programs at their curricular cores; these two prongs, also among STC's 2007 professionalization initiatives, have reached critical development stages and will be launched in the

[5] Available online at http://www.bls.gov/oco/ocos319.htm

near future. These initiatives are the Body of Knowledge and certification of technical writers or communicators. The Body of Knowledge, according to the STC website, will help to define the field as well as identify fundamental principles, knowledge, and tools needed to be successful as a technical communicator.[6] Certification, according to the STC's most recent press release, provides two key benefits to technical writers: "First, certification establishes a solid foundation for the legitimacy and economic contribution of technical communication. Second, certified practitioners can clearly demonstrate their expertise as technical communicators, greatly enhancing their value in the marketplace."[7] Both of these actions—certification and developing a body of knowledge—are still in early stages, but, if they are successfully implemented, their effects could ripple strongly through PTSC program curricula. How programs will respond and adapt to these changes and how these changes will affect the market are still unknowns, but their potential impacts could be monumental.

## The Professoriate, the PhD, and the Market

If the trajectories we noted in this chapter are shifting the foundations on which we base the education and resulting employment of practitioners in our field, then have our programs shifted doctoral student training as well to meet new needs and requirements for success? Are we asking them to be active learners and teachers who value learner-centered instruction and practice self-reflection? Are we educating doctoral students to employ the five habits of mind when considering technologies and their instruction? Are we providing them with opportunities to explore cultural differences and consider the effects of localization and globalization in their own work so that they can teach others through their experiences? Are we empowering them to succeed in the shifting and uncertain financial situations that many higher education institutions are currently experiencing? In essence, these questions require faculty in doctoral-granting programs to prepare students with more than domain knowledge in PTSC; their preparation must also include metadiscussions about what it means to be a professor in today's market.

Among these discussions should certainly be consideration of what it means to conduct a job search for a professorial position. Insights from search committee chairs who sought to fill technical writing positions at university and colleges across the United States in 2007 suggest that many doctoral graduates need more training in how to engage in job search activities. The survey, the second in a longitudinal study of the technical communication job market, found that the number of positions advertised in PTSC was stable, with 117 positions

[6] Available online at http://stcbok.editme.com

[7] Available online at http://notebook.stc.org/a-monumental-day-dawns-for-technical-communicators-certification/

advertised in 2002 and 118 advertised in 2007. Similarly, in both 2002 and 2007, over three-quarters of the available positions were located in English departments, and approximately 60% of the advertisements in both surveys sought technical and professional communication as the primary emphasis for job applicants. The 2007 survey, however, did reveal signs of dissatisfaction in graduate training, especially in job search skills. In contrast to the 2002 survey, researchers discovered that many search committee chairs were disappointed with the professional preparation of many on-campus applicants. Chairs noted that applicants seemed unaware that research-intensive institutions expected applications to have publications prior to graduation and that search committees valued job talks highly during on-campus interviews. One search committee chair noted that

> on the whole, the search committee was disappointed with candidates who did not seem ready to be on the market. Their job talks were weak, and candidates didn't ask good questions about the department or acted like they weren't interested in the position for which they were interviewing.

Another chair said she was "shocked by the poor quality of the teaching demonstrations. Candidates did not put enough time and effort into job talks or teaching demonstrations. The majority of candidates were mediocre or worse in job talk, a deciding factor in selection." A third common complaint of the search committee chairs related to practical expertise, which many applicants did not appear to have: Although all institutions expected students to be able to discuss their dissertations, including topics and methodologies employed, even more important to search committees was the applicants' ability to describe practical applications of their work in terms that everyone, including undergraduate students and faculty members in other departmental emphases, could understand. The need to be able to explain how one's research applies in practical situations was a frequent theme in the search chairs' comments. These findings suggest that doctoral-granting programs have more work to do in preparing their students to acquire their first professional positions.

A related issue concerns research funding, as described in the earlier section on higher education financing, and doctoral program training, a concern that has serious implications for graduate programs. Recently, the administration at Michigan State University (MSU) and Texas Tech University (TTU) both began requiring faculty to include grant-seeking abilities as a criterion in selecting all faculty applicants. In other words, if you don't know how to "make rain," do not apply. The PTSC programs at MSU and TTU have been given four $25,000 doctoral fellowships for the upcoming 4 years. To fulfill the requirements for the fellowship, doctoral students have to submit at least one outside grant proposal within 2 years of receiving the fellowship. When the market requires grant-making abilities and competencies, we would seem to be obliged to teach students how to write winning proposals. Are our programs preparing doctoral students for this work? And do our programs, especially those in English or language and literature departments, have the infrastructure to support this grant-making and grant-getting work? We are skeptical that most do. If we are correct, newly minted doctorates find themselves in a bind. They will be required to obtain grant funding, but their departments are not yet prepared to help them administer or conduct such work.

Preparing students to respond to the changing needs of the market as well as to adapt to changing institutional requirements is primarily the responsibility of the training institutions, but several of our professional organizations have responded to these challenges with support of their own for graduate students. Begun at the start of the 2000s, two CPTSC initiatives have been designed to strengthen membership, in general, and graduate student, in particular, professional development. The first of these initiatives—the CPTSC Research Grant initiative—has grown from awarding one research grant of $500 per year to awarding three $1,500 grants per year since its beginning. Of the 29 members who have received research grants, 6 recipients, or approximately 20%, have been graduate students. More recently, the CPTSC established the Graduate Student Diversity Scholarship, intended to increase the diversity of CPTSC membership. The scholarship waives the recipient's registration for the CPTSC conference and provides the recipient with $500 for travel and other conference-related expenses. Similarly, at its 2009 conference, the Association of Teachers of Technical Writing (ATTW) included research and professional development workshops for graduate students. Initiatives and workshops like these provide graduate students, doctoral students in particular, with valuable opportunities to promote their professional development in what it means to be a member of the professoriate.

Although it is clear that our programs and our professional organizations are addressing many needs of doctoral students and through them, addressing the needs of the programs in which they will teach, there is much work yet to be done. Clearly, some of this work is descriptive research that tells us more about the doctoral programs as well as the successes and failures of those trained in these programs. For example, we need to know how well calibrated are matriculation, graduation, and placement rates of doctoral students, and whether we are adequately preparing them for good positions. We might also need to consider our expectations for these graduates. Does every PhD recipient need to be a researcher and publisher? The University of Delaware, as an example, has designated several positions on the writing faculty as teaching professorships: non–tenure track, continuing positions assigned a full workload of teaching four courses per semester. Teaching professors are paid on par with tenure-track faculty. They have full voting and participation rights, enjoy full benefits, and have career tracks with opportunities for promotion and increments that parallel the tenure-track lines. Are our doctoral programs designed to meet the needs of faculty such as these? If not, what changes are needed to meet these needs? More programmatic research can help us to answer these questions and to understand more fully how well we are training the future professoriate for our programs.

## MOVING FORWARD

Hindsight provides us with the ability to look to the past to identify trends that have influenced and sometimes determined our work as program directors and as leaders of the CPTSC, an organization primarily concerned with the health and welfare of PTSC programs. Less clear, however, are the future directions toward which these trends will lead us. For our programs, the dominant future issue may well be how to move as fast as the world moves, or as two of us have recently discussed elsewhere (Maylath et al., 2010), how to think of our programs intellectually so that we do not continually chase after trends, which we fear happens too much. Perhaps our wish will always remain a dream not quite realized; however, we suspect that all readers will agree that PTSC programs will be positioned best if we can track the trajectories of trends before us and thoughtfully respond to the issues that they present us.

## REFERENCES

Andrews, D. C., & Henze, B. (2009). Teaching professional writing to American students in a study abroad program. *Business Communication Quarterly, 72*(1), 5-20.

Ambrose, S., Bridges, M., DiePietro, M., Lovett, M., & Norman, M. (2010). *How learning works: Seven research-based principles for smart teaching.* San Francisco, CA: Jossey-Bass.

American Academy of Arts and Sciences. (n.d.). *Humanities resource center online: Humanities indicators.* Retrieved from http://www.humanitiesindicators.org/humanities Data.aspx

Association of American Colleges and Universities (AACU). (2007). *College learning for the new global century. A report from the National Leadership Council for Liberal Education and America's Promise.* Washington, DC. Retrieved March 23, 2010, from www.aacu.org/leap/documents/GlobalCentury_final.pdf

Bernstein, E. (2006, April 12). Colleges impose restrictions on study-abroad programs. *The Wall Street Journal*, p. D1.

Bransford, J. D., Brown, A. L., & Cocking, R. R. (2000). *How people learn*. Washington, DC: National Academy Press.

Carroll, J. M., & Rosson, M. B. (1992). Getting around the task-artifact cycle: How to make claims and design by scenario. *ACM Transactions on Information Systems, 10,* 181–212.

Courant, P. N., & Knepp, M. (2008). *Budgeting with the UB model at the University of Michigan*. Retrieved from www.provost.umich.edu/budgeting/ub_model.pdf

DeVoss, D. N., Cushman, E., & Grabill, J. T. (2005). Infrastructure and composing: The *when* of new-media writing. *College Composition and Communication, 57,* 14–44.

Dobrin, D. (1983). What's technical about technical writing? In P. V. Anderson, R. J. Brockmann, & C. R. Miller (Eds.), *New essays in technical and scientific communication: Research, theory, practice* (pp. 227–250). Amityville, NY: Baywood.

Dubinksy, J. M. (2004). Becoming user-centered, reflective practitioners. In J. M. Dubinksy (Ed.), *Teaching technical communication* (pp. 1–10). Boston, MA: Bedford/St. Martin.

Faris, M., Lundin, R. W., & Selber, S. A. (2010). *Remixing reading contexts: e-Reading in academic settings.* Paper presented at the 61st annual convention of the Conference on College Composition and Communication, Louisville, KY.

Fink, L. D. (2003). *Creating significant learning experiences.* San Francisco, CA: Jossey-Bass.

Florida, R. (2010). *The great reset: How new ways of living and working drive post-crash prosperity.* New York, NY: Harper.

Friedman, T. L. (2005). *The world is flat.* New York, NY: Farrar, Straus and Giroux.

Gnecchi, M., Maylath, B. A., Mousten, B., Scarpa, F., & Vandepitte, S. (2011). Field convergence: Merging roles of technical writers and technical translators. *IEEE-Transactions on Professional Communication, 54.*

Holmes, R. (2008). *The age of wonder.* London, UK: Harper.

Huba, M. E., & Freed, J. E. (2000). *Learner-centered assessment on college campuses.* Needham Heights, MA: Allyn & Bacon.

Institute of International Education. (n.d.). *Open doors 2006: U.S. students studying abroad.* Retrieved April 11, 2007, from http://opendoors.iienetwork.org/page/89252/

International Society for the Scholarship of Teaching and Learning (ISSOTL). (2009). *ISSOTL history.* Retrieved from http://www.issotl.org/history.html#founding

Johns, E., & Draper, N. (2010, May 26). Thousands of Minnesotans might not graduate. *Star Tribune.* Retrieved from http://www.startribune.com/local/94865234.html?elr=KArksUUUoDEy3LGDiO7aiU

Johnson-Eilola, J. (2005). *Datacloud: Toward a new theory of online work.* Cresskill, NJ: Hampton.

Maylath, B. A., Grabill, J. T., & Gurak, L. J. (2010). Intellectual fit and programmatic power: Organizational profiles of four professional/technical/scientific communication programs. *Technical Communication Quarterly, 19,* 262–280.

Maylath, B. A., & Thrush, E. A. (2000). Café, thé ou lait? Training technical communicators to manage translation and localization. In P. Hager & H. J. Scheiber (Eds.), *Managing global communication in science and technology* (pp. 233–254). New York, NY: John Wiley & Sons.

Newfield, C. (2009). Ending the budget wars: Funding the humanities during a crisis in higher education. *Profession, 15,* 270–284.

Selber, S. A. (2009). Institutional dimensions of academic computing. *College Composition and Communication, 61,* 10–34.

Starke-Meyerring, D., & Wilson, M. (Eds.). (2008). *Designing global learning environments: Visionary partnerships, policies, and pedagogies.* Rotterdam, The Netherlands: Sense.

Tagg, J. (2003). *The learning paradigm college.* Bolton, MA: Anker.

Walvoord, B. E., & Anderson, V. J. (2010). *Effective grading: A tool for learning and assessment in college* (2nd ed.). San Francisco, CA: Jossey-Bass.

Watson, R. N. (2010, March 21). The humanities really do produce a profit. *The Chronicle of Higher Education.* Retrieved from http://chronicle.com/article/The-Humanities-Really-Do-Pr/64740/Pr /64740/

Wieman, C. (2007, September/October). *Why not try a scientific approach to science education?* Retrieved from http://www.changemag.org/Archives/BackIssues/September-October2007/full-scientific-approach.html

Wieman, C., Perkins, K., & Gilbert, S. (2010, March/April). Transforming science education at large research universities: A case study in progress. *Change, 42*(2), 7–14.

Wood, B. S. (2009). *Lecture-free teaching: A learning partnership between science educators and their students.* Arlington, VA: National Science Teachers Association.

CHAPTER 3

# Program Assessment: A Passion and Palimpsest

***Nancy W. Coppola***

"Accountability and assessment of outcomes." When faculty hear these words, they often think of a blunt mechanism that will stifle their academic freedom and ability to evaluate student's critical thinking skills. Faculty-administrators may be even more apprehensive as they marshal defiant instructors for a task viewed as an unrewarded time-sink. In fact, faculty reluctance is the single biggest impediment to the adoption of systematic assessment of learning outcomes (Katz, 2010). And, as a colleague in technical communication program leadership wrote to me about her experiences in the bureaucracy of upper-level assessment meetings, "I found it so frustrating that we were never talking about program/department culture and assessment practices at the same meetings, as if culture were a reward and assessment a punishment to all involved" (personal communication, February 18, 2009).

Indeed, program assessment imposed by the federal government, accreditation agencies, or university administrators without faculty ownership will not work. Likewise, there is no systematized metaparadigm or static methodology for good program assessment. Writing assessment and, by extension, writing program assessment must be site-based and locally controlled, assessment researchers Brian Huot (2002), Edward M. White (2005), and Robert Broad (2003) tell us. According to assessment historian Norbert Elliot (2005), "The history of writing assessment is a series of case studies, linked inextricably to individual program developers and unique design circumstances" (p. 347). This chapter, then, is one such case study, the story of one program developer whose initial perspective on all program leadership activities was naïvely myopic, yet who continues to develop a vision that is increasingly panoptic.

The book prospectus for the present volume notes the dearth of published research in program administration with the exception of assessment: "In this case alone, program administrators have benefited greatly from colleagues' expertise, which is often gained through lengthy experimentation and trial and error" (Bridgeford, Kitalong, & Williamson, 2008, pp. 3–4). Notable too is the fact that most of the assessment scholarship is presented in the expected tradition of research studies that review the literature, describe the methodology, and report results. This chapter attempts to describe the personal story of self-reflection and knowledge—not only knowledge gained by empirical research but also by wisdom of practice. Fortunately, given memory's malleability, I have recorded much of this history. Following epistemology studies of Polyani (1966) and the knowledge spiral of Nonaka and Takeuchi (1995), my early tacit knowledge, rooted in study and teaching of technical communication, was converted to explicit knowledge through articulation in presentations and publications, which then triggered assessment-driven instruction in our programmatic communities.

As I began to reconstruct my story, I realized that my experiences mirror the assessment movement itself in higher education. Therefore, this chapter, as palimpsest, will recount my personal narrative contextualized in the trajectory of our field: the late 1980s, moving away from standardized testing; early 1990s, moving toward community; mid-1990s, establishing a graduate program as research site; 2001, enduring our first program review; 2004, engaging our first program assessment; and 2010, mapping institutional values to program outcomes.

## LATE 1980s—MOVING AWAY FROM STANDARDIZED TESTING

In the late 1980s, when I was a special lecturer at New Jersey Institute of Technology (NJIT), teaching four sections of our undergraduate service course, I was invited to Princeton, New Jersey, to take part in a focus group of technical writing instructors. The Educational Testing Service (ETS) gathered teachers to review its Defense Activity for Non-Traditional Education Support (DANTES) Subject Standardized Test of Technical Writing. These tests were administered to those in the armed forces and civilians, who could, with a passing score, then apply for college credit. The test content, although based on the topics commonly taught in the classroom, was operationalized through multiple-choice questions and sentence-correction items. We were asked, for example, if the content in a question set about the genre of instruction represented material we taught in our courses. Of course, we taught instruction writing in our courses, but I created interactive modules with Tinkertoys rather than a multiple-choice test item that asked students to identify imperative voice as the correct distinguishing characteristic of the genre. I was conflicted by my view of pedagogy, grounded in the theories of learning advanced by Kenneth Bruffee (1984, 1986), who wrote persuasively about the social construction of knowledge and collaborative learning, and the prevailing paradigm of multiple-choice tests to assess knowledge of technical writing.

My conflict mirrored the larger debates going on in mainstream writing assessment and in theories of technical writing. In 1989, Egon G. Guba and Yvonne S. Lincoln challenged the scientific paradigm of inquiry that standardized tests represented and offered a revolutionary constructivist approach that seems strikingly contemporary in hindsight. They urged a sociopolitical process that involved collaboration and consensual values for evaluation, localization for individual contexts, and a Socratic dialogue in which stakeholders both teach and learn from the assessment. Led by Carolyn Miller and her landmark article (1979), technical communication theory was also emerging from its 1970s positivist epistemology, in which technical communication was perceived as a tool for transmitting knowledge from one mind to another in clear, objective, and neutral language. In constructing a brief history of writing assessment in the introduction of *Assessment in Technical and Professional Communication*, Allen (2010) recounted the struggle of social constructivism during the late 1980s and across the 1990s to become the key element in authentic practice of teaching and assessing writing.

## EARLY 1990s—
## MOVING TOWARD COMMUNITY

I was fortunate to be welcomed into a culture of assessment in 1991 when I returned to NJIT from a leave of absence. Robert E. Lynch, professor of English at NJIT, was deeply involved in writing assessment as chair of the New Jersey Basic Skills Council. Working with the New Jersey Basic Skills Placement Test, his research with holistic scoring in large-scale essay assessment helped determine that New Jersey students were successfully placed in appropriate college courses. Norbert Elliot, then associate professor of English, brought his knowledge and practice of assessing outcomes to NJIT from Texas A&M-Commerce, where he was coordinator of composition and director of the North East Texas Writing Project. Their work with basic skills assessment was short-lived as the New Jersey governor, in concert with national attacks on open admission policies at many colleges and universities, which gave rise to the need for basic writing, simply dissolved the State's Board of Education and its testing system. Two decades later, composition scholars again predict the need for basic writing and other compensatory programs due to the recent influx of increasing numbers of nontraditional students to the nation's colleges and universities (Mutnick, 2011; Otte & Mlynarczyk, 2010).

Lynch and Elliot brought their sense of accountability and stakeholder responsiveness to NJIT's undergraduate service course in technical writing. Together with Margaret Kilduff (1994), they developed a formal writing assessment program that was deployed over 3 years. In the first year, 1988–1989, they field-tested and refined a common curriculum they had developed; in the second and third years (1989–1991), they developed a formal scoring methodology. Using a portfolio method to assess education outcomes, their study addressed fundamental issues of sound assessment—reliability and validation.

When I joined the research group in 1991, I came with practitioner's knowledge and scholarship of technical communication but no experience in assessment. Elliot and Lynch were patient and generous mentors, not only welcoming me into the research group but also piling my desk with the important literature of assessment. Therefore I came to know Pat Belanoff and Peter Elbow (1986), Hunter M. Breland (Breland, Camp, & Jones, 1987), Roberta Camp (1982), and Robert J. Jones, in addition to assessment scholars noted earlier in this chapter.

We continued to assess outcomes of students in the junior-level technical writing course (ENG 352) using a scoring rubric and 4-point scale. Toward the end of each semester, students conferred with their instructors to select their strongest writing samples from their portfolio of work. Then the undergraduate technical writing instructors met for a 4-hour reading session. Six preselected sample clusters were presented to the readers. Readers began by reviewing a cluster illustrating a high level of student performance; its merits were described in terms of the scoring rubric. Readers then reviewed a second cluster, a sample demonstrating low performance. The final four samples were read in silence, and the readers nominated a score for each. Extensive discussion then followed. Once the group of technical writing instructors reached consensus on the scoring rubric, the papers from the current semester were then scored. Each cluster, or subset of the student's portfolio, was read and scored by two readers, neither of whom was the student's instructor. To continue the objectivity provided by this measure, scores were masked so that the score given by the first reader could not influence the second reader. Clusters with discrepant scores were then given a third reading, although the readers did not know the previous readers' scores. This procedure was followed consistently in the 3 years of the study and in subsequent years.

This conceptual framework of assessing learning outcomes allowed me to theorize how socially constructed models of pedagogy would be implemented in the technical communication service course (Coppola, 1999). In my study of the cluster scoring, I found evidence of collaboration between and among students and teachers, student as active participant in the learning process, meaning dependent on social and cultural contexts, feedback and iteration, and a workshop environment. Equally important to sound assessment practices, we found portfolio reading (and the concomitant reliability and validation) as measures of our ability to reach consensus, to work together in a community, to exercise careful thought, and to apply fair judgment. The collaborative model of research continues today with new partners (Johnson & Elliot, 2010) and with data that show the patterns of cluster scores remaining stable across time. Although the technical writing service course is not strictly a program, the focus on outcomes assessment in a conceptual framework suggests programmatic thinking, according to Carter, Anson, and Miller (2003); therefore, I include this event as part of my program administration history.

## MID-1990s—ESTABLISHING A GRADUATE PROGRAM AS RESEARCH SITE

My introductory experiences as a program director are likely consistent with others: I became a program director at the same time I became an entry-level assistant professor, my first significant moment in program administration was serendipitous, and, although I had no experience in administration, I also had an open playing field on which to write our future. In 1993, we proposed a graduate program in professional and technical communication, the first technical communication program at NJIT and graduate program in the Humanities Department. We offered our first classes in fall 1994 with two full-time students. In that first semester, we produced a live local TV broadcast, "Technical Communication Today," in which each faculty member made a hands-on presentation in an applicable skill area for weekly broadcasts. In fall 1995, with 10 students, I offered our first online course, then called a telecourse. Students rented videotapes of the "Technical Communication Today" series, and our course was conducted with a conferencing system platform developed at NJIT—Electronic Information Exchange System (EIES). When this course won the Cable Award for Programming Excellence (CAPE Award) for the Best Series/Telecourse in the education category, I was hooked on online teaching. Faculty interests in online teaching and new technologies were reinforced by student demand for more online courses. By 1999, our master's program was completely available online as well as on campus. But as student demand shifted, we moved to a completely online program in 2002.

As a public research university, NJIT required an active research agenda in addition to my administrative responsibilities. The Master's in Professional and Technical Communication would become my administrative and pedagogical home as well as site for research. Fortunately, I was a co-principal investigator on an Alfred P. Sloan Foundation grant—From Virtual Classroom to Virtual University: Institutionalizing Asynchronous Learning Networks at NJIT. Working with distinguished scholar Starr Roxanne Hiltz, we studied pedagogical roles in becoming a virtual professor. Our qualitative research of faculty and student relationships on building trust in virtual teams has been widely published. This research informed my teaching and brought quality assurance to the program, a practice I continue today.

## 2001—ENDURING OUR FIRST PROGRAM REVIEW EXPERIENCE

I have subtitled this chapter "A Passion and Palimpsest." "Passion" suggests compelling emotions and these, indeed, have marked my experiences with program assessment.

In 2001, NJIT initiated an internal review process, as scheduled by the provost and the Review Committee for Department and Program Assessment, which functioned as a program audit. Criteria for the review included faculty qualifications and scholarly productivity, research dollars, teaching loads, cost per credit hour, enrollment, retention and graduation rates, time to degree, degrees awarded, courses offered, and so forth. We followed the process guidelines, similar to those we had often used to help our university prepare for review by the Accreditation Board for Engineering and Technology (ABET) and the Middle States Commission on Higher Education (MSCHE). The evaluation procedure was conducted and reported in routine fashion: we presented the curriculum vitae of faculty, the number of courses taught and grades earned, the student demographics, and our plans for future growth. On the surface, the process was essentially a routine audit: the product was a lengthy document written by committee; the process was a staged review performed in a conference room; and the outcome was a final memo certifying that we could continue our recruitment efforts. Behind the scenes, it had all the political struggle for power that characterizes turf wars among deans in a closed room. I was in that room to witness the supportive engagement of the science and liberal arts dean who found the program innovative, the ire of the architecture dean who wanted to own any territory earmarked "media," and the sharp criticism from the computer science dean who judged the program feeble.

But at that time, unaware of the political machinations, I simply felt the sting of rebuke as my personal failure. I had fallen victim to a judgmental stance that Jimmie Killingsworth (1990) warned program administrators against: "The first thing we need to do ourselves is deliberately to abandon questions of success and failure [in program assessment]. . . . Instead of worrying about what our deans need to hear, for a moment, let's think about what *we* need" (p. 38). I also felt a vivid sense of injustice to our students, our faculty, and the program itself. With all the data collected in the report, we knew nothing about how or if our program was meeting its advertised program goals. What exactly were our graduate students able to do? What were they supposed to know? And how were we to measure educational quality in an online environment?

Those beginning first years at the turn of the 21st century were heady times for technical communication. We had survived the dot-com bust and were buoyed by the broad shift from industrial to information economy in which technical communication is a key commodity. We were the knowledge workers, introduced by Peter Drucker (1959), who work with our heads and produce ideas, information, and knowledge. We saw impressive growth in the number of academic programs and courses and an academic job market in which there were several job opportunities for every qualified candidate.

If you were to search the literature in 2001 for a framework on which to build a better model for program review, you would have found an overwhelming diversity of opinions. A highly conceptual framework was constructed by Johndan Johnson-Eilola and Stuart Selber (2001) for graduate education in technical communication. Their model invites academics to situate modes of analysis along a three-dimensional axis of thinking, doing, and teaching to better understand these interrelationships and thereby any activity or artifact. A pragmatic approach could be found in the Council for Programs in Technical and Scientific Communication (CPTSC) Guidelines for Program Self Study, a detailed guide that would lead program representatives through a structured self-study process before an onsite review by an external reviewer. But the CPTSC self-study would yield the program performance data we had already collected in the program audit. Too, we had Jo Allen's 1993 important bibliography of technical communication assessment, the first work that moved us beyond basic discussions to examine deeper concerns, purposes, and criteria for assessment. Characteristic of Allen's assessment scholarship, she gave us questions for self-discovery that would lead to an individual assessment strategy. But I needed answers, not questions.

I remember the next important event in my narrative journey quite vividly because it was what Robert Johnson (2009) called, in his personal reflection and rhetorical mediation on programs in scientific and technical communication, "a kairotic moment." In an open conference room among our offices, I attempted to sketch out what I knew about our master's program and what I needed to know. On legal-sized paper, I drew the familiar five pillars of quality that the Sloan-C Foundation developed to guide continuous quality improvement in online learning: (a) learning effectiveness, (b) cost-effectiveness and institutional commitment, (c) access, (d) faculty satisfaction, and (e) student satisfaction. Although cost-effectiveness was outside our purview, we knew we had institutional commitment through continued funding of our program and dedicated faculty lines. Our traditional course evaluations and student surveys told us that students were satisfied with their program and the accessibility of the learning platform. Benchmarking our program against others had given us a more precise understanding of the instruction and curriculum. And faculty willingly gave their time to teaching and advising graduate students. But what did we know about learning effectiveness?

Enter my colleague Norbert Elliot, who, also a visual problem solver, took up his own paper and began to draw a column titled "predictor variables (X)," with five arrows pointing to the dependent variable (Y)—MSPTC program goals. How were these five predictor variables (X), we asked, related to the outcome (Y) of the graduate program to educate students and thus prepare them to be working professionals? And what would be the artifact, or performance, that would allow the students to demonstrate their abilities? The performance, we decided, would be the "Web sensible" portfolios described by Kathleen Blake Yancey (2004) as student products that not only inhabit digital space but also exploit the digital medium. What we could not easily decide, however, were the student outcomes on which to base our assessment. Here we were in unchartered water because of the lack of research on assessment of graduate programs in technical communication. With few standards for disciplinary knowledge and no empirically based body of knowledge for technical communication, we looked to the literature to find core traits. Two studies informed our assessment criteria of eight core competencies: a survey of core competencies by David Dayton and Stephen Bernhardt (2004) and an annotated bibliography of essentials works in the field compiled by Gerald Alred (2003). This was obviously a convenience sampling that in retrospect omitted many earlier significant projects by other scholars.

## 2004—ENGAGING OUR FIRST PROGRAM ASSESSMENT EXPERIENCE

We encountered faculty objections when we first drafted our matrix of student outcomes and asked our program instructors to review these criteria and particularize them by describing how a critical skill might be taught in their program course. We were right on target, according to Broad (2003), who found that dissent, conflict, and initial failure are normal activities in intellectual life after he spent a year listening to assessment conversations on student writing among members of a university's English department. An instructor teaching visual design (one of our core competencies) told us that visual literacy is an emerging construct that lacked substantial criteria validation and a consistent vocabulary for talking about the visual. Fortunately, we had Tiffany Craft Portewig's (2004) taxonomy of visual language to help us understand and articulate the cognitive process of designing visuals and a visual message. We went back to the drawing board and developed program-specific skills to be addressed in our courses.

During our first formative assessment, at the beginning of the spring semester 2004, we distributed our matrix of core competencies, now detailed with a column of descriptors that particularized the competency in terms of how each is addressed in our courses. Program faculty met in a computer lab with a list of student portfolio URLs to undertake a collaborative, formative review. At the end of the session, instructor comments surrounding each of the eight competencies were summarized and sent by e-mail to each of the students. Comments on the formative assessment addressed strengths ("He's the best student in my class right now. Very professional, very diligent. One of my assignments, he did twice. Learning Flash wasn't enough for him, so he did a digital video of his daughter.") and afforded areas for improvement ("The student does high quality work but does not take part in online discussion. I always had the feeling he was absent, he just wasn't there."). The tone of the assessment was both encouraging and positive.

Here again, we met with faculty resistance. One instructor balked at the process, stating that he was busy, he had just given each student a written detailed evaluation of their graded work, and being judged this way might be unfair to the students. We countered by agreeing that assessment is complex, time-consuming, and that there is never ever a good time to do it no matter how hard we plan. We reminded the instructor that grading is an individualistic act, while assessment involves interaction with others so that communal, consensus observations can be made. The formative assessment would let each graduate student know just what we expected in our program and how the student was doing on the path to mastery. Students also gain a holistic sense of their overall performance in the program.

At the close of the spring 2004 semester, faculty met again to conduct a summative review within a quantitative framework; in this instance, faculty used a Likert-scale rubric to score individual student ePortfolios. Following a calibration session in which we reached consensus on the various levels of student work according to a 6-point Likert scale, the instructors each independently read and scored an assigned number of ePortfolios. They read the ePortfolios to analyze each of the eight competencies as these competencies were manifested within student work. An analytic evaluation was completed for each of the competencies and a holistic evaluation for the overall score; each ePortfolio received two independent readings. These summative assessments provided information for further improvement of both student ability and the program itself.

In our assessment, we had to make difficult programmatic decisions when criteria consistently showed that we were not meeting goals. We developed a two-semester required course sequence, replacing a single course, to fully prepare students for communication in a visual world. In the early years, our summative assessments showed a lack of interreader agreement on the interpersonal communication and collaboration scores. The ability of students to demonstrate interpersonal communication and collaboration—process-oriented, transactional constructs—is difficult in an ePortfolio that is a product-oriented event. We instituted reflective statements in the portfolios that asked students to describe, for example, how their actions as group leader in a collaborative project demonstrated interpersonal communication strengths. By 2006, technology had caught up with our needs, and instructors brought podcasting and wikis to their courses, which students then brought forward into their ePortfolios.

My escalating enthusiasm for program assessment was buoyed by a cycle of conference presentations and peer-reviewed publications. Our study had the following conclusions: the variables captured by the ePortfolios were internally consistent; the ePortfolios were able to be read reliably; the students were able to meet or exceed our performance expectations; and the core competencies, reliably measured, are associated with criterion variables such as course grades and cumulative grade-point average. It was time to reach out beyond our institutional walls; a funded CPTSC research project (Coppola & Elliot, 2006) allowed us to create a public website for assessment research and to perform a cross-institutional assessment. Working with colleagues at Texas Tech University—Miles Kimball, Joyce Locke Carter, and Thomas Barker—is an enduring highlight of my career. In spring 2006, we partnered with the Master of Arts in Technical Communication (MATC) team at Texas Tech to share assessment strategies. The field test produced a collegial exchange and a combined set of core competencies. Reading again the transcript from our conversations in TTU's MOO, a multiple-user text-based conferencing system, I laugh as I read the inserted smiles, claps, and hand waves that sparked our exchange, and I am warmed by our camaraderie.

My narrative time-block in this section—2004 to 2009—reflects higher education's push for accountability and for learning outcomes based on what contemporary college graduates need to know and be able to do. *College Learning for the New Global Century*, a 2007 report by the National Leadership Council for Liberal Education and America's Promise, developed principles that call for a far-reaching shift in the focus of schooling from accumulating course credits to building real-world capabilities. Employers, too, were concerned about skill levels of college graduates for advancement beyond entry-level jobs, a 2008 survey shows:

> When it comes to the assessment practices that employers trust to indicate a graduate's level of knowledge and potential to succeed in the job world, employers dismiss tests of general content knowledge in favor of assessments of real-world and applied-learning approach. (Peter D. Hart Research Associates, 2008)

Surveyed employers found that supervised internships, community projects, ePortfolios, and senior projects develop important real-world skills as well as for employers to evaluate graduates' readiness for the workplace. At the same time, accreditation agencies were beginning to require outcomes-based assessment, a point that I will further develop in the next section.

With the two graphs we had sketched out in 2002 of pillars and predictor variables, we had mapped a way forward. The model of outcomes-based program assessment informs our shared research agenda that continues today. Our publications provided administrators with meaningful evidence of authentic learning. With a change in the senior administration, and three provosts later, the culture of assessment began to change at NJIT.

## 2010—MAPPING INSTITUTIONAL VALUES TO PROGRAM OUTCOMES

The program review audit that I described in the 2001 section of this chapter faltered and was discontinued because departments were reluctant to participate in a process through which inefficient or underutilized programs could be terminated. As well, the process offered little that would help instructors design a gap analysis—a determination of what students learned, where students had fallen short, and where curricular revision was justified. Simply put, the old program-review process was not robust enough to provide a framework to assess and monitor student learning outcomes. The new program-review process implemented in 2010 changed the focus from a contentious audit to a forward-looking model: one that helps the university determine that students have knowledge, skills, and competencies consistent with NJIT goals. The new program review process is enabling NJIT to determine that, at graduation, students have achieved goals appropriate to New Jersey's Science and Technology Research University.

In August 2010, Provost Ian Gatley selected our graduate program to fast-track the new program review process, to provide user feedback on documenting outcomes, and to complete our report by mid-September for the official launch to the entire university. Because of programs-based outcomes assessment, we were expected to develop an exemplary model. In preparing the self-study, the program and the academic unit's program review team director were asked to keep two goals of this process in mind: to ensure that NJIT is offering quality academic programs (e.g., has appropriate educational objectives and is accomplishing those objectives as evidenced by student learning outcomes assessment) and to demonstrate continuous improvement in the delivery of the educational curriculum.

NJIT's Strategic Plan (2010–2015) sought deeper mission penetration throughout the university and its colleges and programs. Therefore, the outline for the program review document required a description of the program's mission and goals and an explanation of its alignment with the university mission and goals. Our MS PTC program report first demonstrated enhancement of university mission and goals and then extended institutional values through programmatic values into objectives and outcomes that result, ultimately, in the determination of learning. In our tabulated data, we showed educational objectives as a fourth column that extended from Institutional Goals to MS PTC Program Goals. To demonstrate the integration of MS PTC Program Goals and Education Objectives in courses, another table showed the articulation of objectives as core competencies. The final table demonstrated articulation of our programmatic core competencies into our core courses. Through these mapping iterations, we were able to show how our university's unique mission and goals shaped our programmatic curriculum and situated our program in its institutional culture.

I will let the letter from Provost Gatley tell the end of my tale (May 2, 2011):

> We are writing to let you know that your report regarding the Master's in Professional and Technical Communication has been approved by the Sub-Committee on Assessment. As part of the new NJIT Program Review process, your report on the design, assessment, and total quality improvement aspects of the program you direct was exemplary. Indeed, we are now using your report as a model for others in our five year review cycle.
>
> In addition, your research undertaken with your student Phoebe DelBoccio was important in the development of our assessment protocol for distance learning. The rubric and review process you provided will allow us to ensure that distance learning courses offered in the core of a program will meet the high standards of quality that we expect from traditionally taught courses. In that your Master's is offered entirely online, you again set a high standard for program assessment at NJIT.

Another provost—Jo Allen (2010)—now President of the Meredith College—advised grounding programmatic assessment in institutional values:

> As such, [institutional values] are highly useful in our conversations and directions for assessment, especially in assessing connections between programs and the larger institutional context. In fact these values can and should serve as a guide or template for the kinds of qualities that faculty should consider when developing and assessing their curriculum. (p. 40)

Her excellent 2010 chapter in *Assessment in Technical and Professional Communication* provided a series of templates and rubrics to help program coordinators explore this assessment practice.

Allen's chapter appeared in the book section "Assessment in the Widest View." This heading seems to give credence to the introductory description of my chapter as the story of one program developer whose initial perspective of all program leadership activities was naïvely myopic yet who continues to develop a vision that is increasingly panoptic. Tenacity has probably been the key attribute in my personal narrative, fueled by an emotional response to my early rejection (as I inferred it) in the first program review episode described. Although this probably tells a good deal about my personality, it also portends what may be required of a true program assessment champion—someone who has an emotional connection—someone with a passion for program assessment.

## REFERENCES

Alred, G. J. (2003, November). Essential works on technical communication. *Technical Communication, 50*(4), 585–616.

Allen J. (1993). The role(s) of assessment in technical communication: A review of the literature. *Technical Communication Quarterly, 2,* 365–388.

Allen, J. (2004). The impact of student learning outcomes assessment on technical and professional communication programs. *Technical Communication Quarterly, 13*(1), 93–108.

Allen, J. (2010). Mapping institutional values and the technical communication curriculum: A strategy for grounding assessment. *Assessment in Technical and Professional Communication* (pp. 39–56). Amityville, NY: Baywood.

Belanoff, P., & Elbow, P. (1986). Using portfolios to increase collaboration and community in a writing program. *Writing Program Administration, 9*(3), 27–40.

Breland, H. M., Camp, R., & Jones, R. (1987). *Assessing writing skill.* New York, NY: College Entrance Examination Board.

Bridgeford, T., Kitalong, K. S., & Williamson, B. (2008). *Sharing our intellectual traces: Narrative reflections from administrators of professional, technical, scientific communication programs.* Book prospectus for the Baywood Technical Communication Series.

Broad, R. (2003). *What we really value: Beyond rubrics in teaching and assessing writing.* Logan, UT: Utah State University Press.

Bruffee, K. (1984). Collaborative learning and "the conversation of mankind." *College English, 46,* 635-652.

Bruffee, K. (1986). Social construction, language, and the authority of knowledge: A bibliographic essay. *College English, 48*(8), 773–790.

Camp, R. (1982). *Proposal for writing portfolio project: Phases I and II and progress report for writing portfolio project, phase I.* Princeton, NJ: Educational Testing Service.

Carter, M., Anson, C. M., & Miller, C. R. (2003). Assessing technical writing in institutional contexts: Using outcomes-based assessment for programmatic thinking. *Technical Communication Quarterly, 12*(1), 101–114.

Coppola, N. W. (1999). Setting the discourse community: Tasks and assessment for the new technical communication service course. *Technical Communication Quarterly, 8*(3), 249–267.

Coppola, N. W., & Elliot, N. (2006). *A community research model for assessment of programs in technical and scientific communication* (CPTSC Research Grant Proposal). Newark, NJ: New Jersey Institute of Technology.

Coppola, N. W., Elliot, N., Barker, T., Locke, C., & Kimball, M. (2007, March). *Re(building) technical communication as a research discipline: A community model for program assessment.* Presented at the Association of Teacher of Technical Writing, New York, NY.

Dayton, D., & Bernhardt, S. A. (2004). Results of a survey of ATTW members, 2003. *Technical Communication Quarterly, 13*(1), 13–43.

Drucker, P. (1959). *Landmarks of tomorrow.* New York, NY: Harper.

Elbow, P., & Belanoff, P. (1999). *A community of writers: A workshop course in writing* (3rd ed.). New York, NY: McGraw-Hill.

Elliot, N. (2005). On a scale: A social history of writing assessment in America. *Studies in Composition and Rhetoric, 3.* New York, NY: Peter Lang.

Elliot, N., Kilduff, M., & Lynch, R. (1994). The assessment of technical writing: A case study. *Journal of Technical Writing and Communication, 24*(1), 19–36.

Guba, E. G., & Lincoln, Y. S. (1989). *Fourth generation evaluation.* Newbury Park, CA: Sage. Retrieved from http://books.google.com/books?hl=en&lr=&id=k_zxEUst46UC&oi=fnd&pg=PA7&dq=guba+and+lincoln+1989&ots=_Xb9whgRJV&sig=4YNzUC-OuoT3OPntviytj4ZGFDw#v=onepage&q=guba%20and%20lincoln%201989&f=false

Huot, B. (2002). *(Re)articulating writing assessment for teaching and learning.* Logan, UT: Utah State University Press.

Johnson, C. S., & Elliot, N. (2010). Undergraduate technical writing assessment: A model. *Programmatic Perspectives, 2*(2), 110–151.

Johnson, R. R. (2009). Trajectories, kairos, and tulips: A personal reflection and meditation on programs in rhetoric, technical, professional, and scientific communication. *Programmatic Perspectives, 1*(1), 45–58.

Johnson-Eilola, J., & Selber, S. (2001). Sketching a framework for graduate education in technical communication. *Technical Communication Quarterly, 10*(4), 403–435.

Katz, S. N. (2010, October). Beyond crude measurement & consumerism. *Academe, 96*(5), 16-20.

Killingsworth, J. (1990). *Evaluating service programs in scientific and technical communication: How qualitative research can help us.* San Diego, CA: Council for Programs in Technical and Scientific Communication.

Miller, C. (1979). A humanistic rationale for technical writing. *College English, 40*(6), 610–617.

Moore, J. C. (2005). The Sloan Consortium quality framework and the five pillars. *The Sloan Consortium.* Retrieved from http://sloanconsortium.org/publications/books/qualityframework.pdf

Mutnick, D. (2011). Review: Basic writing and the future of higher education. *College English, 73*(3), 322–336.

Nonaka, I., & Takeuchi, H. (1995). *The knowledge-creating company: How Japanese companies create the dynamics of innovation.* New York, NY: Oxford.

Otte, G., & Mlynarczyk, R. W. (2010). *Basic writing.* West Lafayette, IN: Parlor.

Peter D. Hart Research Associates. (2008). *How should colleges assess and improve student learning: Employers' views on the accountability challenge.* Washington, DC.

Polanyi, M. (1966). *The tacit dimension.* London, UK: Routledge.

Portewig, T. C. (2004). Making sense of the visual in technical communication: A visual literacy approach to pedagogy. *Journal of Technical Writing and Communication, 32,* 31–42.

White, E. M. (2005). The scoring of writing portfolios: Phase 2. *College Composition and Communication, 56,* 581–600.

Yancey, K. B. (2004). *Postmodernism, palimpsest, and portfolios: Theoretical issues in the representation of student work.* Champaign, IL: National Council of Teachers of English.

CHAPTER 4

# Establishing an Outcomes Statement for Technical Communication

***K. Alex Ilyasova and Tracy Bridgeford***

## INTRODUCTION

In *The Outcomes Book: Debate and Consensus after the WPA Outcomes Statement*, Barry Maid (2005) adapted the first-year composition program outcomes to technical communication programs. He expertly adapts the language of the WPA Statement to fit technical communication courses and programs by adding, modifying, or deleting specific words or phrases, staying well within the categories established by the Council of Writing Program Administrators (WPA). He articulates well his reasons for any changes from the Statement, using arguments clearly within the boundaries of technical communication concerns. Although certainly a very good start, we see this effort as indicative of the problem in the technical communication field, namely, that we categorize, interpret, and explain our work from a standpoint of first-year composition. It's easy to see why and how naturally this positioning occurs because most of us receive pedagogical training for first-year composition. But when assigned technical communication courses, these same graduate students don't necessarily receive the same level and depth of training for technical communication. Despite an interest in and scholarly focus of technical communication programs, it is commonplace to require MA and PhD students to prepare and teach in first-year composition programs, most obviously due to the economic needs of departments for (cheap) composition teachers. Second, the perspective of technical communication as a subfield of composition establishes a vertical hierarchy with composition as primary for any teaching of writing. Finally, the likeliest academic explanation for a lack of training in teaching technical communication is the academic background held by the previous two generations, namely, literature (first generation) and rhetoric/composition (second generation). Consequently, until recently, little preparation and training has been provided for technical communication instruction.

Although economic factors still influence decisions about staffing and training, we see some changes shifting attention away from composition, including more and specialized knowledge of technical communication theories and practices, which means that our body of knowledge is no longer simply lessons about form and style. We are also no longer comfortable with labels that identify our field as a subdiscipline of composition because what we know about technical communication can no longer fit into that small a space. The last decade has seen demands for more space in the general education writing curriculum, a shift in perspective, as the first-year experience may not necessarily still focus exclusively on composition. We see local responses to the needs and to programs as the largest influence leading to these changes as directors consider (as they always have) the local conditions in which they administer their programs. Perhaps with the third, and current, generation of technical communication PhDs, this lack of training can change, because they have been able to focus more exclusively on technical communication theories and practices. If the same emphasis was placed on technical communication pedagogy, what would that training look like? We think part of answering this question, in a way that would aid shifting the focus to technical communication theories and practices, is the creation of our own WPA statement and outcomes; one not adapted from a composition-technical communication as subfield perspective. Our goal in this chapter is to offer up such an outcomes statement, invite discussion about it, and ultimately call for our professional organizations, specifically the CPTSC, to adopt an outcomes statement specifically for and appropriately identified by technical communication theories and practices.

We decided to propose the creation of a technical communication outcomes statement out of a shared concern that a composition standpoint continually hinders efforts in the field to define technical communication, its theories, its practices, and its identity. This hindrance is evident in our ongoing conversations to define the field and to identify the field's value—our value. Even as we are calling for the creation of technical communication–specific outcomes statement, we are well aware that this call is also in response to seeing the similarities between what motivated the creation of the original WPA Outcomes Statement and our current issues in technical communication. In other words, we are optimistic that creating a technical communication outcomes statement will help us address some of the same issues that the WPA Outcomes Statement did for composition. For example, the WPA Outcomes Statement was intended to speak to those inside and outside the discipline about the essence of composition programs, to speak to these stakeholders in a language that showed the status of the field, and to be a step toward professionalizing the field of first-year composition at a time when the professional work of flourishing graduate programs seemed to be denied or dismissed. In "The Origins of the Outcomes Statement," Edward White (2005) explained that the

> Outcomes Statement gives us a sophisticated and mature view of what college writing programs seek to do, demonstrating why these writing courses belong in the college curriculum, usually at the center of general education, and what other faculty should expect their students to have learned in these courses. (p. 7)

Like the original WPA Outcomes Statement, we intend the technical communication outcomes statement to primarily address an audience of writing program administrators and writing teachers, while also supplying information to other stakeholders in technical communication—students, other administrators, parents, legislators, the public at-large—who have some right to know (Rhodes, Peckham, Bergmann, & Condon, 2005). As a result, the technical communication outcomes statement and the discussion/debate we hope ensues serves to progress the ongoing conversations about defining the field and the field's value to the forefront by establishing a basis for equity—for what technical communication teachers share in common and a measure for what students should be able to do after going through a technical communication program.

In what follows, we first discuss what we have identified as an essence of technical communication—*techne*—the theoretical, practical, and pedagogical aspects of making. This is, for us, in the broadest sense, what we share in common as technical communication teachers and program administrators. Starting with what we consider to be our field's enterprise, *techne* allows us to frame an identity for technical communication that speaks to who we are and what we do. Building on this discussion of value, in the next part, we discuss what we considered when composing a technical communication outcomes statement, or what we have termed the TC-WPA Outcomes Statement. We discuss in more detail the stakeholders we considered and two specific factors that shaped the construction of the statement: the students served by technical communication programs and the decision to create "outcomes" versus "standards." In the last two sections, we discuss our rationale for the categories and the individual statements we composed as well as the importance of adapting the outcomes to individual programs by illustrating how each of us has done just that. We conclude with an invitation to continue this work and a specific call for our professional organization to take on this task.

## *TECHNE* AS ENTERPRISE

The outcomes offered in this chapter grow out of the values teachers and scholars in technical communication have emphasized in conversations during the last several decades, including, though not limited to, critical thinking, situated learning, collaboration, user analysis, genre conventions, and social construction of knowledge. One conversation has resurfaced in the midst of current discussions about legitimacy, identity, and power (Faber, 2002; Kynell-Hunt & Savage, 2003, 2004; Savage, 1999): What is technical communication?[1] Generally, scholars in and outside the field have attempted to define this illusive term, first to distinguish it from literature as a possible means for increasing awareness of its role in an academic department or an organization. More current conversations seek to establish a definition that can signify the value of technical communication and by extension the technical communicator such as Corey Wick's (2000) effort to connect technical communication to knowledge management, a distinctively 1990s term. These conversations focus on the value technical communicators contribute to a project or product, to emphasize the ways that kind of work adds economic value to their role in an organization as a means for increasing their credibility and worth on the job (Carliner, 1997; Wick, 2000). More broadly, a definition could embody the disciplinary values with legitimacy by establishing a body of knowledge that certifies the practice as well as the practitioner (STC BOK). In academia, a definition contributes to a program's ethos in ways that clearly speak to what the program is about and what it can offer students. Whether intended for academia or industry, these conversations sustain the well-known rationalistic versus humanistic argument (or instrumental versus rhetorical) that has been going on for at least three decades (see e.g., Miller, 1979).

Each definition created to date has covered well the object of technical communication work (e.g., "technology"; Dobrin, 1989), the contexts in which technical communication occurs (Sullivan, 1998), and the practices technical communicators follow (Miller, 1991, *praxis*; Miller, 1989, *phronesis*). But we'd like to reorient the discussion. Like Jo Allen, who in 1990 argued that we should "abandon the search" to define technical writing, we agree with her that such a definition

> would have to be an extensive and flexible definition that will represent the complexities and delicate balances of content, purpose, style, format, and all the other components of successful technical writing—not only the components of current technical-writing practices, but those of the twenty-first century as well. (p. 76)

It would take away from us, we think, the very thing that makes us special—flexibility. However, focusing on enterprise as a definition can better enable the

[1] Our intention here is not to summarize efforts to define technical communication or even to offer our own definition, but to acknowledge the scholars who have provided comprehensive treatments of this conversation (see, e.g., Allen, 1990).

flexibility and adaptability needed to meet local demands because the local supersedes any scholarly arguments that attempt to create a universal understanding of what we do and what we teach. In defining the field (the work) in terms of its enterprise (what we value), we can position ourselves in ways that can help us better establish universal outcomes. The inability to create an agreed-upon definition will not take away from our identity or what we value as a community of practice—our enterprise.

From Etienne Wenger's (1998) perspective, an enterprise is a joint endeavor "not in that everybody believes the same thing or agrees with everything, but in that it is communally negotiated" (p. 78). The negotiation of the enterprise is what enables the community to develop coherence, a condition that does not necessarily have to be explicitly stated. Most communities of practice, in Wenger's theory of social learning, are not officially organized communities. Rather, they tend to grow organically from an interest in a particular practice, its value, and a commitment to learning more about it to improve it (see, e.g., Julian Orr, 1996, description of Xerox service repairpersons' formation of a community of practice). For technical communication, we believe that with or without a distinct definition, our coherence as a community still conducts itself effectively, and the value of our practice is evident to members and growing more evident to outsiders. Some problems early scholars in the field discussed at length (see, e.g., the first decade of CPTSC proceedings) focused most obviously when discussing hiring and tenure constraints. Although some programs still might exhibit problems with power and identity, the crux of the problems seems to have ebbed somewhat as the number of programs and tenure-track faculty grew. This is not to say that the battle for identity is not ongoing, but we do want to point out that our shared enterprise—*techne*—marks us as special, with or without a shared definition of what we do and who we are. And what we value, what is shared throughout the scholarly literature, is a focus on *techne*. Whether we call it *techne* or not, this value—our enterprise—is what keeps us together, willing to continue learning and negotiating its meaning.

*Techne*, a classical rhetoric concept most often associated with Aristotle, has been so modernized with such terms as *technical*, *technique*, and *technology* that its original meaning has been lost. We see this as an unfortunate circumstance because the richness of the classical sense of *techne* is worth exploring, starting with Dobrin (1989), who defined "technical writing as writing that accommodates the user to technology," creating an unbreakable impression of our enterprise as technology. By connecting *techne* to the "military-industrial complex," the corporate world (see, e.g., Dale Sullivan, 1991) reduced *techne* to technical, creating an equally unfortunate bond to efficiency of work, or what Katz (1992) called an "ethics of expediency." Carolyn R. Miller (1989) set up *techne* in opposition to praxis by focusing on conduct and reflection of conduct. In a special issue of *Technical Communication Quarterly* (TCQ) on *Techne*, Ryan Moeller and Ken McAllister (2002) did a thorough job of summarizing these and other arguments (such as Doheny-Farina, 1992) that reduce *techne* to an "instrumental" concept. Criticizing this instrumental perspective, Moeller and McAlister and Jay Gordon (2002) in the same special issue of TCQ provided a different understanding of *techne*, which moves it away from the rationalistic, modernized, and stripped away version offered both in and outside the field of technical communication.

Emphasizing this difference, Joseph Dunne (1993) defined *techne* as "the kind of knowledge possessed by an expert maker," giving the maker a "clear conception of the why and wherefore, the how and with-what of the making process." This conception is what gives the maker a sense of "secure mastery" (p. 9). The competence afforded by *techne*, then, makes different the outcomes of a task, whether a physical object or cognitive process.[2] *Techne*, Dunne said, is not the same thing as production; it does not create "anything tangible at all" (as cited in Moeller & McAllister, 2002), but in fact, it happens *a priori* to production. The competency displayed by the maker is developed over time, through trial and error and through and in negotiation of a practice, and is the enterprise we undertake when teaching and researching technical communication. Hoping to start a dialog about the theory and practice of technical communication reminiscent of the classical discussion of Plato and Aristotle's discussions, Jay Gordon (2002) explained, using Roochnik's (1998) Homeric conception of *techne*: *techne* has a determinate task, either tangible or intangible; it is productive, but may not produce a material object; it must be applied for expertise to develop; it must benefit the community; and it relies on general principles (pp. 150–151). *Techne*, Dunne (1997) would agree, belongs in the realm of general principles.

As our enterprise, *techne* emphasizes the value of making that leads to production, and making and production, to paraphrase Jimmie Killingsworth (1992), are always influenced by local concerns. Programmatic decisions and pedagogical methods are modified from the general theories garnered from scholarly articles, hallway conversations, or conference discussions. To make something is to apply what can be used from local resources. To develop a program is to demonstrate local needs and desires. What works at a research institution just won't work at a 4-year college whose central mission is teaching. The enterprise of *techne*, as it relates to our field, is localized within the university and campus enterprises where we live and work and invest our identities, and is modified as necessary. We continually walk a tightrope of concerns, weighing the feasibility of global concerns of identity, autonomy, and power with local realities of resources, location (both the city as well as the department), and ownership (such

[2] See Joseph Dunne's *Back to the Rough Ground* (1993), Janet Atwill's [QA: There is no reference for Atwill.] *Reclaiming Rhetorica* (1998), and Jeffrey Walker (2000) for a rich discussion of *techne*.

as with words like *communication*), just to name a few. Although these localized situations afford the most opportunities available for program development and pedagogy, we continue to learn from each other, regardless of location. But that one area where we come together as a community of practice is *techne*—we develop, teach, and implement ways of making and in doing so, continue to learn from each other. *Techne,* in other words, makes possible these outcomes.

## THINKING THROUGH THE TC-WPA OUTCOMES STATEMENT

The TC-WPA outcomes statement aims to make explicit these shared practices, what our field values, and what we expect students to learn in our programs. Additionally, as with the WPA Outcomes Statement, we intend for the TC-WPA outcomes statement to assist program administrators with doing their jobs in terms of assessment, curricular and programmatic development, and outreach to the variety of stakeholders across their campuses and in their local communities.

As mentioned in the introduction, the following stakeholders are our primary and secondary audiences: adjuncts, instructors, graduate students, and of course, program directors (primary audience); and deans, students, parents, legislators, and teachers in other disciplines (secondary audience). Keeping them in mind, we wanted to strike a balance between writing in our professional language (which would speak to our disciplinary identity and keeping the complexity of meaning associated with terms like *production* and *user*) and making the document accessible to stakeholders outside the discipline. As Rhodes et al. (2005) explained when they considered language for the WPA Outcomes Statement,

> Professional language, characterized by words like *rhetoric*, *genre*, and *conventions* (and *register*), is useful to people who have grown used to a common set of associations, including the historical uses of the terms. But, to others, it smacks of snotty language people use to show that they understand because they are on the in—and of course people who don't understand are on the out. (p. 15)

Whether this perspective is a correct interpretation or not, it is something we also admitted as a possibility when we composed the document. Striking the right balance meant that we considered the length of the document, the words we chose for the categories, the content and organization of the individual statements, and when to use terms that were particularly reflective of our field and that marked its identity.

The statement should then articulate what we expect students to have learned by the end of their undergraduate experiences in a technical writing/communication class or a technical communication program. This articulation, we think, presented a challenge because most of our programs serve two different populations of students: the general population (i.e., those taking only the services courses, usually in their second year) and the technical writing/communication majors (i.e., those taking the service courses and/or courses specific to the major throughout their entire undergraduate career). To address these populations, we needed a statement that, as Kathy Yancey (2005) stated, had a "'gateway in' approach—one focused on the exit point of the curriculum, regardless of where that might be" for students (p. 218). This gateway-in approach focuses on moving students into the rest of the curriculum (as is the case for the general population of students, and we would add, moving technical communication students through the technical communication curriculum and into their next experience), whether in industry or graduate school. In addressing both populations, we follow the example of the original WPA statement by keeping some of the outcomes more general, as well as including terminology specific to our field. The latter is reflected in our word choices, the organization of the outcomes, and the categories we created. For example, we have technology and collaboration as their own categories, but include ethics in each category. The intent was to address both populations of students who move through our programs and describe the common knowledges, skills, and attitudes we expect them to learn at their varying exit points.

Lastly, we paid attention to the distinction between "outcomes" and "standards." Based on the work of the original WPA Outcomes Statement group, we agree that creating outcome statements and not standards was more useful for the broad goals we've established, and mainly because outcomes allow us to *not* have to agree on the single best way to achieve them. In other words, we can reach consensus about what matters in the curriculum without having to reach consensus about how to get there. As Kathleen Yancy (2005) stated,

> *Outcomes* provides another way of talking about and understanding curricular work . . . through thinking about what is it that we want students *to know, to understand*, and *to do* at the conclusion of a course, a program, a major, we begin to articulate our expectations: or, *outcomes*. (p. 21)

Moreover, and this aspect is the truly exciting part for us, outcomes-based assessment, as Yancy (2005) discussed, addresses both teaching and learning, individual as well as program assessment, and tends to look at both product and process. Outcomes, as she stated, "provide a kind of curricular stability without being very invasive—which in part explains their appeal" (p. 22). All these aspects seem incredibly useful in helping TC-WPAs do their jobs better with assessment and curriculum and program development. Because we have then, through outcomes-based assessment, the ability to consider thoughtfully both product and process, as well as provide stability in curriculum but flexibility in curricular specifics like teaching strategies, pedagogies, and the kinds of assignments, we can hopefully avoid having the argument that reaching consensus here limits us or ignores the important tensions that have allowed our field to advance. We have no intentions of being prescriptive; we are instead, like the WPA Outcomes Statement, divorcing outcomes from standards in the hopes that the TC-WPA outcomes meet the needs of the wide range of institutions we represent.

## TC-WPA OUTCOMES STATEMENT

The TC-WPA outcomes statement then is not merely a compilation or summary of what we believe currently takes place. Rather, the following statement articulates what technical communication program administrators and teachers nationwide develop, teach, and implement, that is, their *techne*, the knowledge and skills learned from practice, research, and theory. Keeping these aspects in mind, the questions that guided us were the following:

- What are the aspects of technical communication that we value in terms of learning, making, and production?
- What knowledge, understanding, and skills do students acquire as a function of participating in a technical communications program?
- At the conclusion of the program, what do students know? What can they do?

Here then is the TC-WPA outcomes statement we composed, with the categories and the individual outcomes that grew out of those questions.

### Research (Critical Thinking, Reading, and Writing)

Students will show they can:

- Use research methods for gathering information;
- Understand production as including gathering, evaluating, analyzing, and synthesizing appropriate primary and secondary sources;
- Understand the relationships among language, human behavior, and power;
- Understand the main features and uses of writing in their fields;
- Address reader/user expectations;
- Interpret findings and articulate results;
- Understand the potential for misrepresenting information and misleading readers/users;
- Produce appropriate and ethical graphics for displaying research data and findings; and
- Document the resources as defined by the field.

## Practices and Processes

Students will show they can:

- Conduct user/reader analysis;
- Focus on a defined purpose;
- Meet the needs of the readers/users;
- Respond effectively and ethically to different rhetorical situations;
- Understand writing as a collaborative and iterative process of research, discussion, negotiation, and editing;
- Plan in stages; and
- Develop flexible strategies and behaviors for generating, revising, editing, and proofreading.

## Knowledge of Conventions and Genres

Students will show they can:

- Write in multiple genres;
- Learn common formats for different genres in the context of a discipline;
- Understand how each genre shapes content and usability;
- Demonstrate knowledge of genre conventions with regard to format, design, structure and paragraphing to style; and
- Demonstrate control of tone, syntax, grammar, punctuation, and spelling.

## Collaborative Learning

Students will show they can:

- Understand how to participate and communicate in a community;
- Understand the collaborative and social aspects of writing processes and document invention;
- Negotiate meaning in works-in-progress collaboratively;
- Integrate ethical ideas and content from various stakeholders;
- Critique technical documents from a variety of sources;
- Balance the advantages of relying on others with the responsibility of doing their part; and
- Demonstrate a sense of responsibility for completion of work both individually and collaboratively.

## Technology

Students will show they can:

- Choose from a variety of technologies to address a specific situation and a range of reader/user needs;

- Understand the ethics and role of technology and the human practices associated with it in their field;
- Demonstrate a critical perspective of technology, its uses, and contexts;
- Analyze technology as a physical tool, as a metaphorical text, and as a socially constructed system; and
- Demonstrate competence with various software for writing, editing, and designing.

The five categories represent the general principles we have programmatically and pedagogically in common; ones we have created from our joint enterprise—*techne*. In composing the TC-WPA outcomes statement, we kept a number of aspects from the original WPA Outcomes Statement, for example, the attention to research, process, and knowledge of conventions. However, in crafting the outcomes statement to better reflect our field and what we value, we added the categories of Collaborative Learning and Technology, and distributed the individual outcomes previously under Rhetorical Knowledge, thus, eliminating the category altogether.

With regard to the Technology category, we agree with Maid (2005) and the differences he articulated about including technology in the outcomes for technical communication. He stated,

> Frankly, if I thought students in my program could expect to work as writers who only used standard word processing software, I would have hesitated to include the technology outcome in my revised criteria. However, since my students will be expected to work not only in multiple genres but in multiple media, the issue of understanding the rhetorical choice and effect of technology becomes crucial. (p. 144)

But we would like to extend Maid's focus on technology by including a separate category that highlights more explicitly the role it plays in our field. The importance of technological knowledge and skills for technical communicators to program curriculum and development is most evident in job descriptions (see Faigley & Miller, 1982). Of course, as Johndan Johnson-Eilola (1996) reminded us in the mid-1990s, it is not a good idea to design curriculum as "wish-lists of skills to industry" (p. 247). Although this warning is accurate in spirit, the cultivating of specific skills is necessary and appropriate for technical communication programs. However, we need to carefully consider how to incorporate technology in the classroom and what outcomes are expected. Most of us probably agree that teaching technology means teaching the skills for learning technology as opposed to teaching one specific skill or software. This category engages multiple aspects of using, situating, and conceptualizing technology in the classroom as well as ethical sensibilities that provide the boundaries for understanding its framework and its context (see, e.g., Katz, 1992). Even when antagonistic, we can't ignore the role technology plays in our pedagogical and programmatic decisions.

Similarly, the focus on collaboration, whether in document production, team communication, or project development, is also most evident in job descriptions and equally appropriate as an explicit area of focus in the TC-WPA outcomes statement. Collaborative writing, we know, exists in the professional setting. Faigley and Miller's 1982 survey confirmed this for us long ago. Creating a separate category that focuses on collaboration makes explicit, and again highlights, what we value in our field: a shared responsibility for preparing documents, but also more broadly, addressing directly with the social constructions of knowledge and some of its complexities.

By listing technology and collaborative learning in separate categories, the intention is not to segregate them into a more single, self-important light. Rather, we expect that WPAs will situate their technological and collaborative learning outcomes (as well as the other outcomes in the other categories) in ways that make sense to their departmental needs and local concerns. We expect that the outcomes listed under each category will be applied as appropriate to individual programs regardless of specific terminology or localized circumstances. In the next two sections, we provide two examples of how the TC-WPA outcomes statement was adapted to two different programs.

## Applying TC-WPA Outcomes Statement, Case #1[3]

Three factors influenced the revisions we, as a faculty in the Professional and Technical Writing (PTW) program, made to the TC-WPA outcomes statement, all of them deriving from our understanding of the institutional contexts and needs of the University of Colorado at Colorado Springs (UCCS). The first factor involves our institution's core values and how we applied them as a backdrop for our work in the program. As Jo Allen (2010) noted,

> Such a context—program within institution—may seem obvious, but from my years of experience in assessment and more recently as a provost, few assessment plans or strategies fully take on the institution's core values as a backdrop for their work. (p. 40)

The second factor involved considering what can be used from our institutional resources to develop the program in ways that respond to the local needs and desires of our communities. Finally, our experiences with students in the classroom as well as our training shaped what we wanted to articulate in terms of student learning. In other words, what we wanted was to create a TC-WPA outcomes statement that reflected not only the values of the field but the mission and

[3] See Appendix A for a list of the UCCS PTW outcomes.

values of the university, our communities, and the program. And, because outcomes are part of assessment, we wanted to move the outcomes from what Chris Anson (2010) described as "wishful thinking . . . to concerted efforts to gauge their manifestation (or lack thereof) in the abilities of students at some point in time" and in this particular place (p. 6).

As the saying goes, hindsight is 20/20; reflecting on our institution's values while writing this chapter has helped highlight the importance of articulating exactly where our program was situated in terms of the institutional culture. Our university is a comprehensive baccalaureate institution, focused on providing "unsurpassed, student-centered teaching and learning," while also offering "a selected number of master's and doctoral degree programs" (Vision Statement[4]). The students here tend to stay in the local area after graduating as opposed to moving or going on to graduate school someplace else. The majority are also older, nontraditional students returning to school in the hopes of advancing or switching careers. And most have families and/or work at least part-time while attending classes. And the majority of traditional-aged students are looking to get jobs in the area once they graduate.

It comes as no surprise then that the core values of the institution reflect these demographics. Here are the first four (out of eight) core values of the university:

- attracting excellent faculty, staff and students;
- helping traditional and nontraditional students be academically successful;
- linking the university with the community it serves; and
- aggressively seek[ing] the development of a multicultural campus environment. (UCCS Vision Statement)

It seems obvious, as Allen (2010) stated, that any program in the university would speak directly to these values. As she explained further, "At its simplest, a mission reflects what the university is (research, doctoral, liberal arts)," which in our case is a student-centered teaching university. Additionally, the mission also reflects "just as importantly, what [a university] does (leads, provides, engages, serves, promotes); institutional values suggest its desired outcomes (civic leaders, global learners, citizens of character, industry leaders)," which in our case means serving our communities by helping students achieve academically (p. 40).

Strengthening these connections through both description and assessment is a particularly valuable means for situating a program while also grounding the program's assessment strategy (Allen, 2010, p. 40). In strengthening these connections ourselves, we noticed that our program fits nicely (and now more purposefully) with the core values and mission of the institution. For example, we have always placed as a priority establishing connections with local organizations

[4] See http://www.uccs.edu/~pages/visionandvalues.html

and businesses that, as a result, developed into a heavy focus on community-based projects for students in the program. Additionally, the size of our program (approximately 40–50 PTW majors during any given year) allows us to know them better in terms of their goals for seeking a degree and their nonacademic obligations, such as work schedules and family commitments. Recognizing these connections has thus helped us stabilize and affirm these priorities. As a result, we have a more stable framework to work within when our discussions move to examining where exactly these values manifest themselves in our curriculum and how they can be articulated in the outcomes statement. And, as Allen (2010) identified, "It is in the curriculum and co-curriculum where the two should meet, with those elements being the repository of evidence of delivering on the values" (p. 40).

Moving on from what a university is to what a university does, I now turn to our curriculum. I start with information about the PTW program, which is in the College of Liberal Arts and Sciences and housed in the English Department. The English Department offers an undergraduate degree in English, with the choice of an emphasis in one of the following areas:

- Literature
- Rhetoric and Writing
- Professional and Technical Writing (PTW)
- Teacher Education Preparation (TEP)

The PTW program consists of me, the one tenure-track faculty member who also serves as program director, and six (non–tenure track) full-time instructors. PTW courses are primarily taught in computer-assisted classrooms, where a variety of production software is available to students and faculty. Courses at the 2000 level also include engineering and business majors fulfilling their writing requirements. The 3000- and 4000-level courses consist of English majors who have chosen the PTW emphasis and students completing a minor in PTW. And, from the local communities, three main industries employ professional/technical writers: (a) military or military contractors, (b) technology and software companies, and (c) religious organizations.

We hear quite often from students that they want, as one of them recently said, "an English degree that I can use in the 'real world'." Given this pressure, one of our main concerns when revising the outcomes statement was to strike a balance between teaching specific skills students want and teaching students how to learn any particular skillset. In other words, not designing the curriculum and outcomes to meet "wish-lists of skills to industry" (Johnson-Eilola, 1996, p. 247). While reviewing the category names from the general outcomes statement previously described, it became clear to us rather quickly that leaving the category of Technology as is would conflate learning the tools with learning the skills for learning technology for the stakeholders we were imagining—students, parents, administrators, and community members. As previously stated, most of us probably agree that teaching technology means teaching the skills for learning technology as opposed to teaching one specific skill or software.

What we realized from our discussions and the discussions with a stakeholder outside the program was that the term that made the most sense for us was *technological literacy*. Based on discussions with the provost (who was also at that time still a faculty member in the Education department) about the National Council of Accreditation of Teacher Education (NCATE) guidelines,[5] addressing *technological literacy*, a term used by NCATE and the provost, was a priority. Tangentially, because the English department also serves the needs of the Teacher Education Program (TEP), it seemed important to consider this aspect when revising the outcomes statement for the PTW program. As a whole, in the English department, we often share students, whether because they switch emphases or because they take a certain number of core courses regardless of programmatic emphasis. As a result, considering the learning goals for the TEP majors has relevance for us in the PTW program, not to mention that it aligns with the conversations about technology we have been discussing in the field. Taking all this information into account, we decided to use the term *technological literacy*—the knowledge, skills, and practices students need to be considered technologically literate—and rename that category. In making this small change, we realized how these outcomes resonate beyond the goals of the PTW program to include other stakeholders at the institutional level, like the provost.

These next revisions reflected the goals of another set of stakeholders—the librarians. As Maid (2005) explained, "Ensuring that our students will be able to work with information as part of their research/writing process is something that writing teachers sometimes take for granted" (p. 144). Like Maid, we collaborate with librarians on issues of information literacy every semester. This collaboration includes participating in their ongoing information literacy project, titled LiONiL[6] and encouraging students to take the workshops librarians offer every semester. Our work together has helped us identify the specific terminology we wanted to use in the PTW outcomes statement that aligned with their goals as well. In particular, under the Research category, we added the word "navigate" to the following outcome: Understand production as including gathering, evaluating, analyzing, and synthesizing appropriate primary and secondary sources. This addition calls attention to students' ability to not just

[5] The provost's attention to the NCATE guidelines highlights the term, *technological literacy,* but a current search for that term was inconclusive. More information about NCATE can be found at http://www.ncate.org/Default.aspx

[6] Library Instruction Online for Information Literacy, or LiONiL, is a tutorial developed and offered by the librarians that teaches basic information literacy skills. LiONiL lays the foundation for developing effective university research skills.

gather, evaluate, and synthesize information but also to navigate to the information they actually need amid the various kinds and amounts of information they encounter. Although this one word might seem like a minor change, for us, it pointed out an aspect of the research/writing process important to development as well as an aspect, that is, in our case, also connected to the goals of these additional stakeholders—the librarians.

We made two additional revisions that called attention to specific aspects of production in the categories of research and technology. These revisions addressed the (often unstated) expectations we have around students being skilled in working with multiple media. As Maid (2005) argued, students will be expected to work not only in multiple genres but in multiple media. Keeping this in mind, we added the word "viewers" to outcome statements that originally included only "readers and users." For us, adding "viewers" to the outcomes meant that we would be asking students to recognize the needs and expectations of not only *readers* and *users* of texts, but also *viewers* of various visual formats, such as videos, photos, graphics, and so on. In particular, we would be asking students to pay attention to what *viewers* look at and how they view it. The term *viewer* then helped us to call attention to these more specific aspects of production.

The remaining revisions came from discussing, as Yancy (2005) stated, "the *what* of education . . . what we want students *to know*, *to understand*, and *to do* at the conclusion of a course, a program, a major" (p. 21). By articulating the *what*, the revisions we made involved moving a few outcomes to other categories, taking some out, and rephrasing the beginning of some in a way that made them both teachable and measurable—for us. In other words, the *what* ultimately had to be in a language and in an order *we* would use and recognize. For example, from the Research category, we decided to move the following two outcomes into the Knowledge of Conventions and Genres category:

- Understand the main features and uses of writing in the field
- Document the resources as defined by the field

We removed a handful of outcomes we thought were either redundant or did not represent an aspect of what we do in the language we found accessible. For instance, we took out the following outcome: "Understand the relationships among language, human behavior, and power." Instead, we decided that the outcomes, which included a category specifically for ethical considerations included in all categories, sufficiently addressed this relationship between language, human behavior, and power. Finally, we changed a word here and there to help make the outcomes more teachable and measurable for us. Such changes included changing the verb "understand," in some instances, to more specific, active verbs like "engage," "evaluate," "identify," or "participate."

Altogether these changes illustrate three strengths about the general outcomes statement. First, that what we can do to it, as Maid (2005) suggested, "may be one of the real strengths of the Outcomes Statement. Minor changes, a word here and there, can make the outcomes statement appropriate for courses and programs other than first-year composition" (p. 142). Second, integrating the skills and preferences of any given program into the outcomes statements is relatively easy, and with just some minor revisions, can turn it into a statement that can be used effectively to evaluate the uniqueness of any undergraduate technical communication program. Finally, as Yancey (2005) stated, "These outcomes present what the profession of [technical communication] values in terms of classroom practice. In this way these outcomes become objects of continued inquiry for practitioners and for WPAs" (p. 30). This case illustrates the essence of technical communication values.

## Applying TC-WPA Outcomes Statement, Case #2[7]

Like Alex, I'd like to start with some facts about the University of Nebraska at Omaha (UNO) and the Graduate Certificate in Technical Communication (GCTC) before describing my adaptation of the general outcomes. UNO is a public, metropolitan university of about 15,000 students, which recently received a Community Engagement Classification from the Carnegie Foundation. The student population is a diverse mix of full- and part-time, traditional and non-traditional students. The graduate certificate programs at UNO are meant to assemble existing resources (e.g., faculty and courses) and reengineer them as concentrated areas of study. They are typically 15–18 credits and require some kind of a completion portfolio. The GCTC consists of five courses that build upon the skills and knowledge of each previous course. Courses include Digital Literacies, Technical Communication, Information Design, Technical Editing, and a Technical Communication Capstone. Ideally, the courses should be completed in order; however, because I am the lone specialist and director for the program, I cannot always offer them in sequence. The GCTC requires 15 credits (4 required and 1 elective) and are cross-listed at both curricular and programmatic levels with the School of Communication.

David Sapp (2006) referred to a faculty member in my situation as a "lone ranger" (p. 200), who faces several challenges such as administrative service while untenured, intellectual loneliness, elitism, and taking sides in the age-old battle with literature (pp. 201–202). These challenges aside, part of the attraction of a lone ranger tenure-track position like mine is the autonomy it affords curriculum and programmatic development and decision-making. The GCTC is entirely my creation, my pedagogical and programmatic vision, but it is also my success or my failure, depending on the circumstances. The lone ranger situation I

[7] See Appendix A for a list of the UNO GCTC outcomes.

created has unfortunately led to a myopic programmatic vision. It is not that my efforts have been unsuccessful; enrollment in technical communication classes has increased with each passing year. But the program has settled into a sluggishness of vision because as a lone ranger, my vision is rarely challenged by faculty members in the know, who might, as in other programs with multiple instructors and rotating directors, bring different perspectives to the programmatic vision and implementation. In addition to the lack of challenges, there is a danger that lone administers can operate within a bubble, which can situate them in ways that do not really make them accountable to the community in which they work.

All aspects of the GCTC program at my school are designed with Etienne Wenger's (1991) concept of communities of practice (COP) in mind. A community of practice is a group of people who share a common interest in a particular enterprise and who mutually engage and negotiate the meaning of the community's enterprise to learn more about it and add value to its enterprise.[8] This COP learning theory is embodied within the program goals and outcomes both in the classroom and in the assessment of the required completion portfolio. I designed the GCTC from a COP perspective because I wanted the idea of acting within a community to constantly be at the forefront of recruiting and instruction. I wanted students to always consider their actions as part of a community that defines its identity, practice, and meaning in a specific way and for a specific purpose. As much as I support the general outcomes Alex and I created in this chapter, I am not comfortable communicating them outright as disparate pieces of a knowledge pie that come together through instruction, because the communities that students will join after graduation will not present performance expectations as a nice, tidy package of skillsets (although arguments can be made that a job description performs that function) that somehow work out as a comprehensive identity. Rather, as Wenger described, identity, practice, and meaning are communicated through actions, stories, and negotiations that require interpretative energy and commitment from members. Students will learn how to become a part of a community by developing an identity of participation that enables students to act as whole persons within a particular situatedness.

COPs emphasize situated learning in which members negotiate the meaning of the community's enterprise, what Jean Lave and Etienne Wenger (1991) called *legitimate*[9] *peripheral participation* (LPP). LPP is a process of learning

[8] For a longer discussion about the role COPs can play the technical communication classroom, see Bridgeford (2006).

[9] I do want to point out that Lave and Wenger (1991) do not use the term *legitimacy* in the Lyotardian (1984) sense of the language games involved in constructing and sustaining grand narratives. Rather, they use *legitimacy* in more of a Miller*esque* (1979) sense of how to belong to a community, that is, as a "defining characteristic of ways of belonging" (p. 35). What make the circumstances legitimate are the affordances made available to newcomers that open up opportunities for engagement (i.e., what they mean by peripheral).

that provides newcomers with full access to a COP's resources in ways that "entail learning as an integral constituent" (p. 35) of a community's practices. What attracts me most to Lave and Wenger's way of understanding learning is that it moves the idea of situatedness, the "internalization" of knowledge, to a consideration of learning as "increasing participation in communities of practice," one that concerns the "whole person acting in the world" (as opposed to a passive receiver of knowledge), emphasizing "comprehensive understanding" of a community, its enterprise, and its identity (p. 49). One belongs because one wishes to learn what it means to be a part of that community, which is not to say that every community is a learning community or that everyone seeks to learn how to belong to every community joined. It does mean, however, that when choosing to belong, newcomers are attracted to the identities, practices, and constructions of meaning that distinguished a particular community. I designed and implemented the GCTC with the LPP theory in mind; it is the enterprise of my program.

As the enterprise, LPP drives its outcomes by keeping the concept of community in the forefront and by emphasizing the construction of identity, participation in practice, and the negotiation of meaning. Ideally, I pictured the "community" aspect of the program would involve multiple stakeholders that would include current and former students as well as professional technical communicators. To some extent, this community-building has happened, but mainly only when I instigate a formal occasion. I had hoped it would become more organic and longer-lasting. However, in the classroom, students do develop a cohesive sense of community that demonstrates the COP outcomes in ways that reveal understanding of the work they do as a community, even when submitting individual work.

Although not word-for-word, the general outcomes in the GCTC Alex and I are proposing for the field embody the same values, skillsets, and learning, but are tailored to the language of communities of practice. The GCTC outcomes include the following categories and their parallels to the general outcomes:

- Expectations for Belonging (Research)
- Participation in Practice (Practices and Processes)
- Identification of Stakeholders (Collaboration)
- Reify the Rules Negotiating Meaning (Knowledge of Conventions and Genres)
- Articulation of Meaning
- Reflection on Membership (Ethics)

I list six categories because I separate two of them: knowledge of conventions and genres. Although I believe that development of a critical perspective of technology is important, I do not include it as a separate category; rather, it is incorporated into each of the other outcomes, because the role of technology today is ubiquitous and involved in all activities, for the most part. Because I understand the communities of practice theory as a comprehensive, collaborative endeavor, I use the term *stakeholders* to emphasize the role and value members bring to a community of practice as part of their mutual engagement and their place in the communicative action as well as the role and value of a product's audience, both of which affect the construction of meaning.

I also don't use the term *genre* to refer to the collectively constructed nature of writing because no matter how much we talk about genre as constructed knowledge (Miller, 1984), the term itself tends to point to the shape of the writing (product) more than it does to a community's activities, the context in which work gets done. Again, because I like to keep community up front, I categorize the "Knowledge of Conventions and Genres" as "Reification of the Rules for Negotiating Meaning" and "Interpretation of Meaning" to emphasize knowledge of a community's shared repertoire of practices and the rules that guide members' mutual engagement and negotiation. The repertoire contains the community's idea of competence and expectations for performance, including "routines, words, tools, ways of doing things, stories, gestures, symbols, genres, actions, or concepts that the community has produced or adopted in the course of its existence, and which have become part of its practice" (Miller, 1984, p. 83). This category emphasizes the aggregate nature of the joint enterprise and the various artifacts that emerge from members' mutual engagement.

Like the general outcomes, I do include an ethical outcome in each category. But unlike the general outcomes, I situate ethics prominently under the category of Reflection on Membership. Although I do not like to separate ethics from other activities as if ethical situations only occur with big events with specific outcomes, I do want to make it more obvious with reflection because membership in communities of practice emphasizes accountability to the community for the contribution members make to the meaning of the enterprise, and accountability happens on a daily basis. Wenger (1998) called this kind of contribution an *ownership of meaning*, characterizing the ability of an individual member to negotiate and "contribute to, take responsibility for, and shape the meanings that matter" in mutual engagement with community members (p. 197). Taking ownership is not an effort to win or be the best of the community, but to add value to the meaning of the community's enterprise in engagement with other members and to be open to negotiation of that construction of meaning. The completion portfolio is used to review student's ability to "make use of, affect, control, modify, or in general, assert as [theirs]" the meanings they negotiate (p. 200), both demonstratively in their completion portfolios and verbally in the accompanying cover memos. The ethics of a situation should be apparent in the student's performance as demonstrated in their completion portfolio.

The completion portfolio must demonstrate to me both verbally and visually how students understand the situatedness of their portfolio. Because most students create a job market portfolio, they must include a cover memo that describes the industry, audience, and contexts in which the portfolio will be interpreted.[10] In other words, students must display the identity of the community, its practices, and their ability to negotiate the meaning of the work described in the portfolio in ways that demonstrate how to belong to that community. The nature of my program is focused on how to belong; the portfolio, then, should make obvious the community they seek to join and how they state their ability to add value to its enterprise.

## CONCLUSION

The TC-WPA Outcomes Statement allows us to focus in on what we do in the classroom and what we truly value in our professional practices. Moreover, it helps address the problem in the technical communication field of our work being categorized, interpreted, and explained from a standpoint of first-year composition or from outside the discipline at a global level or outside the department at a local level. Developing this statement is an opportunity to move past this stagnant perception as well as maintain the continual scrutiny and conversation about the growth of our knowledge, what we value, and how we can articulate it. The two narratives illustrate the adaptability of the outcomes to individual programs and theories. The first narrative illustrates how the outcomes can be aligned to help meet the goals of the institution and reflect the institutional culture, while the second narrative revised the outcomes to reflect the theoretical foundation upon which the program was built. In each narrative, the outcomes statement served to create a productive space for discussions about our enterprise—*techne*—at our individual institutions and with all stakeholders. More importantly, however, these cases illustrate the usefulness of the statement in helping us examine and articulate the values and practices beyond local programs.

Both narratives show that we can reach consensus about what matters to us in the curriculum and in our profession without having to reach consensus about how to get there. To reiterate what Kathleen Yancy (2005) stated so eloquently,

> *Outcomes* provides another way of talking about and understanding curricular work . . . through thinking about what is it that we want students *to know*, *to understand*, and *to do* at the conclusion of a course, a program, a major, we begin to articulate our expectations: or, *outcomes*. (p. 21)

This "curricular stability," Yancy continued, enables us to discuss these outcomes "without being very invasive—which in part explains their appeal" (p. 21). We would add that a TC-WPA Outcomes Statement also provides at least some of the professional stability the field needs, creating a step toward professionalization. It speaks to those stakeholders both inside and outside the discipline,

[10] For a sample portfolio, see http://www.unomaha.edu/techcomm/alinder/index.htm

communicating the essence of technical communication programs in a language that clearly shows the maturing status of the field. And, unlike other reform documents or recommended initiatives—like certification, for example—this one (like the WPA Outcomes Statement) incurs no cost, mandates no structural changes, nor requires new practices for students or instructors. In other words, it is easy to implement and to endorse.

However, to implement it and for our professional organizations to endorse it, we need to take the next step. We need to make use of our professional organizations and structures—specifically, CPTSC and ATTW, and our journals and conferences—to announce, distribute, and solicit responses to this TC-WPA Outcomes Statement *during its development*. We want to get this statement "out there" now as an invitation for others to participate in its formation, to try it out, and to respond. Once this collaboration establishes a general agreement among scholars and instructors in the field, we call on CPTSC in particular to publish in *Programmatic Perspectives* for others to adopt the outcomes in a way similar to how the WPA Outcomes Statement was adopted by the WPA organization and posted on its website. And again, we urge the conversation about its adoption to continue in our journals and conferences as well as at the local levels of our communities and institutions. If the response and the success of the WPA Outcomes Statement is any indication of what can happen with our statement, people will start paying attention to us and begin to know we are here as a discipline separate from composition, thus assuring that we do more than simply provide lessons about form and style.

## APPENDIX A: UNIVERSITY OF COLORADO AT COLORADO SPRINGS PROFESSIONAL AND TECHNICAL WRITING PROGRAM OUTCOMES

### Mission and Objectives Statement

The mission of the PTW program is to prepare students to participate critically and ethically in professional and technical communication positions upon graduation. To that end, the PTW curriculum combines courses from technical and professional communication, rhetoric and writing, and literature in order to help students gain the following:

- a theoretical and historical understanding of professional and technical communication;
- professional and practical skills within the field of professional and technical communication (e.g., skills in written and visual communication, technical writing and editing, critical thinking, and oral presentation);

- an understanding of the ethical concerns, responsibilities, and dimensions of the field;
- technological and visual literacy skills (e.g., document design, graphics, computer documentation, electronic editing, and content management applications)
- the ability to work critically and collaboratively to complete projects.

## PTW Outcomes Statements

### Research (Critical Thinking, Reading, and Writing)

Students will show they can

- Use research methods to gather information
- Evaluate, analyze, navigate, and synthesize appropriate primary and secondary sources
- Identify reader/user/viewer expectations Interpret findings and articulate results
- Produce appropriate and ethical text and graphics for displaying research data and findings

### Practices and Processes

Students will show they can

- Conduct user/reader/viewer analysis
- Focus on a defined purpose
- Meet the needs of the readers/users/viewers
- Respond appropriately and ethically to different rhetorical situations
- Understand writing as a collaborative and iterative process of research, discussion, negotiation, writing, and editing
- Manage projects in stages
- Evaluate and use appropriate strategies for production, revision, editing, proofreading, and presenting

### Knowledge of Conventions and Genres

Students will show they can

- Write in multiple genres
- Evaluate ethically how each genre shapes content and usability
- Control such features as tone, syntax, grammar, punctuation, and spelling
- Identify the main features and uses of writing in a specific field
- Document resources as defined by a specific field

## Collaborative Learning

Students will show they can

- Participate collaboratively with others in the iterative process of research, discussion, negotiation, writing, and editing
- Participate and communicate effectively in a community
- Integrate their own ideas with those from various stakeholders
- Balance the advantages of relying on others with the responsibility of doing their parts

## Technological Literacy

Students will show they can

- Critically and ethically choose from a variety of technologies in order to address specific rhetorical situations and a range of readers/users/viewers needs
- Engage in a critical perspective of technology, its uses and contexts
- Analyze technology as a physical tool and as a socially constructed system
- Use various software for writing, editing, and designing

# APPENDIX B: UNIVERSITY OF NEBRASKA AT OMAHA GRADUATE CERTIFICATE IN TECHNICAL COMMUNICATION PROGRAM OUTCOMES

## Expectations for Belonging (*Understanding how to belong to a community*)

Students will show they can

- Understand identity as a display of competence
- Recognize the enterprise of a community of practice
- Understand the significance of the community's enterprise
- Recognize the circumstances and reasons for belonging
- Recognize the rules that define the community's boundaries
- Demonstrate understanding of a community's expectations for performance
- Identify the rhetorical expectations for a community's situation
- Understand the role technology has in belonging to a community
- Understand the role ethics plays in the construction of identity

### Participate in the Practices of a Community of Practice (*Understanding how to participate in the practices of a community*)

Students will show they can

- Demonstrate access to a community's ways of working
- Demonstrate knowledge of the community's repertoire of practices
- Engage in the negotiation of meaning with other members
- Recognize individual contributions in conjunction with other members' contributions
- Articulate knowledge of the rules for participation
- Demonstrate the use of technology for production
- Understand the role ethics plays in the community's repertoire of practices

### Reifying the Rules for Negotiation of Meaning (*Understanding that meaning is negotiated*)

Students will show they can

- Focus attention through genre conventions
- Recognize meaning within a community's boundaries
- Produce points of reference for organizing meaning
- Understand negotiation as a dynamic process
- Interact with members in the shaping of meaning
- Understand how technology is used to shape meaning
- Understand the role ethics plays in the negotiation of meaning

### Identifying Stakeholder (*Understanding member roles in collaboration*)

Students will show they can

- Recognize key participants in the negotiation of meaning
- Understand the various roles involved in the negotiation of meaning
- Invest and sustain an identity respectfully within a community
- Articulate the importance of the roles in the negotiation of meaning
- Understand the role technology plays in collaborative activities
- Understand the ethics of collaboration

### Interpreting Meaning
### (*Understanding how to create meaning*)

Students will show they can

- Create meaning within a community's boundaries
- Adopt other perspectives through imagination
- Achieve full membership status through the construction of meaning
- Demonstrate understanding of the plurality of meaning
- Interact with members in the shaping of meaning
- Understand how technology is used to create meaning
- Understand the ethics of creating abstractions

### Reflecting on Membership
### (*Understanding participation as responsibility*)

- Take responsibility for constructed meaning
- Demonstrate some degree of control over meaning and the ability to negotiate meaning
- Demonstrate an ability to identify with the community's constructions of meaning
- Add value to a community's repertoire of practices
- Understand how technology affects the meaning of the community's enterprise
- Understand the ethical dilemmas members of a community face on a daily basis

## REFERENCES

Allen, J. (1990). The case against defining technical writing. *Journal of Business and Technical Communication, 4,* 68.

Allen, J. (2010). Mapping institutional values and the technical communication curriculum: A strategy for grounding assessment. In M. N. Hundleby & J. Allen (Eds.), *Assessment in technical and professional communication* (pp. 39–56). Amityville, NY: Baywood.

Anson, C. M. (2010). Assessment in action: A Möbius tale. In M. N. Hundleby & J. Allen (Eds.), *Assessment in technical and professional communication* (pp. 3–16). Amityville, NY: Baywood.

Bridgeford, T. (2006). Communities of practice: The shop floor of human capital. In C. L. Self (Ed.), *Resources in technical communication: Outcomes and approaches* (pp. 161–178). Amityville, NY: Baywood.

Carliner, S. (1997). Demonstrating effectiveness and value: A process for evaluating technical communication products and services. *Technical Communication, 44*(3), 252–265.

Dobrin, D. (1989). *Writing and technique*. Urbana, IL: National Council of Teachers of English.

Doheny-Farina, S. (1992). *Rhetoric, innovation, technology: Case studies of technical communication in technology transfers*. Cambridge, MA: MIT.

Dunne, J. (1993). *Back to the rough ground: Practical judgment and the lure of technique*. South Bend, ID: University of Notre Dame Press.

Faber, B. (2002). Professional identities: What is professional about professional communication? *Journal of Business and Technical Communication, 16*(3), 306–337.

Faigley, L., & Miller, T. P. (1982). What we learn from writing on the job. *College English, 44,* 557–569.

Gordon, J. (2002). *Techne* and technical communication: Toward a dialogue. *Technical Communication Quarterly, 11*(2), 129–146.

Johnson-Eilola, J. (1996). Relocating the value of work: Technical communication in a post-industrial age. *Technical Communication Quarterly, 5*(3), 245–270.

Katz, S. B. (1992). The ethics of expediency: Classical rhetoric, technology, and the holocaust. *College English, 54*(3), 255–275.

Killingsworth, M. J. (1992). Developing programs in technical communication: A pragmatic view. In K. Staples & C. Ornatowski (Eds.), *Foundations for teaching technical communication: Theory, practice, and program design*. Greenwich, CT: Ablex.

Kynell-Hunt, T., & Savage, G. J. (Eds.). (2003). *Power and legitimacy in technical communication: The historical and contemporary struggle for professional status* (Vol. I). Amityville, NY: Baywood.

Kynell-Hunt, T., & Savage, G. J. (Eds.). (2004). *Power and legitimacy in technical communication: Strategies for professional status* (Vol. II). Amityville, NY: Baywood.

Lave, J., & Wenger, E. (1991). *Situated learning: Legitimate peripheral participation*. New York, NY: Cambridge University Press.

Lyotard, J. F. (1984). *The postmodern condition: A report on knowledge*. Minneapolis, MN: University of Minnesota Press.

Maid, B. (2005). Using the outcomes statement for technical communication. In S. Harrington, K. Rhodes, R. Overman Fischer, & R. Malenczyk (Eds.), *The outcomes book: Debate and consensus after the WPA outcomes statement* (pp. 139–149). Logan, UT: Utah State University Press.

Miller, C. R. (1979). A humanistic rationale for technical writing. *College English, 40*(6), 610–617.

Miller, C. R. (1984). Genre as social action. *Quarterly Journal of Speech, 70,* 151–167.

Miller, C. R. (1989). What's practical about technical writing? In B. E. Fearing & W. Keats Sparrow (Eds.), *Technical writing: Theory and practice* (pp. 14–26). New York, NY: Modern Language Association of America.

Miller, C. R. (1996). Comments on "instrumental discourse is as humanistic as rhetoric." *Journal of Business and Technical Communication, 10*(4), 482–486.

Miller, T. (1991). Treating professional writing as social *praxis*. *Journal of Advanced Composition, 11*(1), 57–72.

Moeller, R., & McAllister, K. (2002). Playing with *techne*: A propaedeutic for technical communication. *Technical Communication Quarterly, 11*(2), 185–206.

Orr, J. (1996). *Talking about machines: An ethnography of a modern job*. New York, NY: Cornell University Press.

Redish, J., & Ramey, J. (Eds.). (1995). Measuring the value added by professional technical communicators [Special section]. *Technical Communication, 42*(1), 23–94.

Rhodes, K., Peckham, I., Bergmann, L. S., & Condon, W. (2005). In S. Harrington, K. Rhodes, R. Overman Fischer, & R. Malenczyk (Eds.), *The outcomes book: Debate and consensus after the WPA outcomes statement* (pp. 8–17). Logan, UT: Utah State University Press.

Roochnik, D. (1998). *Of art and wisdom: Plato's understanding of techne*. University Park, PA: Pennsylvania State University Press.

Sapp, D. A. (2006). The lone ranger as technical writing program administrator. *Journal of Business and Technical Communication, 10*(2), 200–219.

Savage, G. J. (1999). The process and prospects for professionalizing technical communication. *Journal of Technical Writing and Communication, 29*(4), 355–381.

Sullivan, D. L. (1998). Political-ethical implications of defining technical communication as a practice. *Journal of Advanced Composition, 10*(2), 375–386.

Walker, J. (2000). *Rhetoric and poetics in antiquity*. New York, NY: Oxford University Press.

Wenger, E. (1998). *Communities of practice: Learning, meaning, and identity*. Cambridge, UK: Cambridge University Press.

White, E. M. (2005). The origins of the outcomes statement. In S. Harrington, K. Rhodes, R. Overman Fischer, & R. Malenczyk (Eds.), *The outcomes book: Debate and consensus after the WPA outcomes statement* (pp. 3–8). Logan, UT: Utah State University Press.

Wick, C. (2000). Knowledge management and leadership opportunities for technical communicators. *Technical Communication, 47*(4), 515–529.

Yancey, K. (2005). Standards, outcomes and all that jazz. In S. Harrington, K. Rhodes, R. Overman Fischer, & R. Malenczyk (Eds.), *The outcomes book: Debate and consensus after the WPA outcomes statement* (pp. 18–23). Logan, UT: Utah State University Press.

CHAPTER 5

# A Tale of Trust and *Techne*: Building Relationships and Building Programs

***James Dubinsky***

> Instead of trusting logical science, we must trust persons, namely those who by long acquaintance with their subject have a right to judge. And if we wish ourselves to share in their convictions and the grounds of them, we must follow their history, and learn as they have learned. We must . . . depend on practice and experience more than on reasoning. . . . By following this we may . . . rightly lean upon ourselves, directing ourselves by our own moral or intellectual judgment.
>
> —John Henry Newman

In Chapter 1 of *The Call of Stories,* Robert Coles (1989) described how he learned to become a practicing, competent psychiatrist. During his internship, he had two supervisors, "two senior psychoanalysts who were there to help [him] make sense of what [he] was hearing" (p. 3). One, Dr. Carl Binger, repeatedly told him to "read more in psychiatric literature" (p. 3) and helped him to see that he had a professional language consisting of words such as "treat," "phobia," and "defenses" (p. 4).[1] Dr. Binger continually exhorted him to "formulate the case" (p. 5), to get "a new appraisal of 'the state of [the patients'] psychodynamics'" (p. 8). The other supervisor, Dr. Alfred O. Ludwig, didn't push him to

[1] Many scholars argued that there are different vocabularies and registers of discourse for different communities (see Gates, 1990 and Saville-Troika, 1989, for instance).

"reformulate," appraise, or sum up his position. Instead, he asked him "to tell . . . a story or two" about the patients he was seeing (p. 11). At first, Coles leaned more in the direction of the former, working to set down the patients' problems and solve the cases, embarrassed by the advice to seek stories. However, stymied by reluctance on the part of his patients to talk, he tried this other "method" of asking them to tell stories; the result is worth repeating:

> For the first time in my short career in psychiatry I saw a noticeable and somewhat dramatic change take place in a patient—and not in response to any interpretation or clarification of mine, but merely as a result of a procedural suggestion, as it were: how we might get on, the patient and I. (p. 12)

Coles learned "how to" learn, "a procedural suggestion" that changed the way he helped patients. He further refined this "method" as he retold the stories to Dr. Ludwig and was asked to concentrate on "clinical moments" (p. 14). These "moments" marked when the person revealed key information that would enable the psychiatrist/listener to help. Despite this initial success, Coles continued to resist Dr. Ludwig (he saw him as odd and out of step with those in power), sticking to reformulating and theorizing. Ultimately, however, he came to understand that listening to his patients' stories helped him make discoveries and act on their behalf (p. 22).[2]

Coles's (1989) story as an apprentice doctor is an appropriate example for a discussion of *techne* that I began in my dissertation and continued in several articles or book chapters, focusing on teaching, program building, and citizenship (1998, 2002a, 2002b, 2010). In a manner similar to teachers, Coles too was "bound up in relationships of interdependency" in "a network of other people" (Dunne, 1993, p. 263), exemplifying a *technite* whose goal is to put his patients "as far as [he can] on the road to health" (*Rhetoric* 1.1.1355b10-14).[3] Thus, like teachers, he is someone in that permeable zone between *techne* and *phronesis*. In addition, he relied on narrative methodology to create knowledge (Clandinin, 1993; Clandinin & Connelly, 2004; Eisner, 1998; Mishler, 1990). His text and story are models from which I learn and borrow, and by borrowing his method and following his examples, this chapter is a narrative "argument" to demonstrate the importance of storytelling to the *techne* of teaching and by extension to program building and administration.

[2] In so doing, he learned what Aristotle explained about the concept of *nous* (intellectual virtue acquired by experience) and judgment: "experience has given [those experienced, older people] them an eye to see aright" (Aristotle, 1962, 166-167, *EN* 1143b9-15).

[3] Aristotle used physicians often in his discussion of *techne* as a way to highlight the need for *kairos* and *phronesis*.

## RELYING ON COLES' MODEL

In this chapter, I'll tell two related stories: a version of a three-semester collaboration with Sarah Michaels, a professor in the Department of Management in the School of Business Administration (SBA), with whom I partnered during graduate school and a version of a 12-year partnership with a local nonprofit organization that I've been involved with since arriving at Virginia Tech.[4] These stories provide some of the intellectual, ethical, and personal traces we've been asked by the editors of this volume to re-create. They provide a foundation and framework for the work I did while building and administering our university's program in Professional Writing. Following Coles' (1989) lead, I focus on "clinical moments."[5] By doing so, I argue for the importance of procedural knowledge, relying on what Knoblauch and Brannon (1988) called phenomenological research methods that present teacher-researchers with a field that values languages and stories excluded from other kinds of academic discourse.[6] I demonstrate that "our work must be situated in practice and with practitioners as we try to understand practice, teacher knowledge, and the ways in which teacher knowledge is constructed and expressed in practice" (Clandinin, 1993, p. 178).

I admit that my tales won't necessarily be verifiable or replicable. They do, however, fall under a framework that Mishler (1990) offered that he labeled "validation," the "process(es) through which we make claims for and examine the 'trustworthiness' of reported observations, interpretations, and generalizations" (p. 419). By focusing on "trustworthiness" rather than "truth," Mishler argued that we move the argument to a social world, "constructed in and through our discourse and actions, through praxis" (p. 420). To represent my "trustworthiness," I "lay bare" my "rhetorical intents" and acknowledge that my narrative is a rhetorical construction (Doheny-Farina, 1993, p. 263). Such a move is essential given my argument for teaching and program administration as a *techne* that is both practical and productive knowledge (Dubinsky, 2002a). As practical knowledge, as *phronesis*, there is an ethical

[4] It is a version because it is my story. As many have argued, to include Brodkey, Geertz, and Van Maanen, what the ethnographer/researcher tells is presented from a biased perspective. As such, it is a "tale," a story. In this version, I've changed the names of the participants to protect their identities.

[5] I use this term because I am not telling (nor could I) the entire story of the collaboration. Rather, I chose a few exemplary instances that help me accomplish my goal of highlighting the interaction of theory and practice (see Coles, 1989, Ch. 1).

[6] Such knowledge is essential to the *techne* of teaching. As Heilbrun (1988) showed, we write our lives. See also Clandinin (1986), Clandinin and Connelly ("Studying") (2004), and Britzman (*Practice*) (2003). Gee (1985) argued that the narrative form is essential: "One of the primary ways—perhaps *the* primary way—human beings makes sense of their experience is by casting it in a narrative form" (p. 11).

component involved (Dunne, 1993, p. 252) that "deals with what is just, noble, and good for man" (Aristotle, *NE* 1143a.12.21–22). As productive knowledge, the teacher's (and, by extension, administrator's) work exists for the good of the students. As Aristotle said, it (the work of the *technitai*) exists "essentially for the good of the governed, as we see to be the case in medicine, gymnastics, and the *technai* in general" (*Politics* 3.6.1279a1–2, as cited in Dunne, 1993, p. 325).

## Clinical Moment #1

My work with Dr. Sarah Michaels was part of a larger collaborative project that two graduate students and I had proposed to the dean of the School of Business Administration (SBA) the summer after our second year as graduate students.[7] Nearly finished with our coursework, we had taken methodology courses and tried our hands at qualitative research. Our personal goal was to gain experience working across disciplines, learning about writing by researching, and testing ways to more effectively teach writing in other disciplines and discourse communities. Our proposed objective was to assist the SBA faculty in integrating writing and technology into their capstone courses and provide students with "portable concepts" and "portable skills" (Perkins, 1986, pp. 222–223) that they could take with them into other courses and into the workplace.

We submitted a proposal that was accepted, primarily because, at the time, the SBA was getting feedback from its Business Advisory Council about graduates' lack of effective communication skills, particularly writing skills.[8] Problems such as these are not uncommon. For many years prior to our work and since, scholars have documented that "business and industry continue to report problems in their employees' ability to write effectively" (Tebeaux, 1988, p. 9). One of the means of solving these problems is to establish collaboration between Technical Communication (TC) or Professional Writing (PW) faculty and instructors in other disciplines. Collaboration of this type entails hurdles to overcome, but usually all parties to the collaboration (to include the students) believe the work is worth the effort (Gelinas, Rama, & Skelton, 1997, p. 382). In our case, we didn't promise that we would solve all of their

[7] The project was developed by Dr. R. Johnson, who was one of the faculty members in our department.

[8] There was another issue as well. The SBA had recently been given a considerable amount of money that it had used to build a state-of-the-art technological classroom. At the time of our proposal, few of the professors were using it. We hoped to offer some suggestions for the use of the electronic media to effect more writing/thinking in the classrooms. I was very interested in this issue, having spent the previous year working with the university's Technology Center to integrate listservs and bulletin boards into my classrooms.

problems, but we did promise that by working with a few professors, we would attempt to solve some and, more importantly, lay a foundation that others could build upon.[9]

We began this project along the lines of what Schön (1983) called "technical rationality," expecting to be problem solvers who would select the technical means best suited to particular purposes. As Schön explained, rigorous professional practitioners solve well-formed instrumental problems by applying theory and technique derived from systematic, preferably scientific, knowledge (*Reflective Practitioner*). This is the argument persuasive to those in the administration, and it made sense for us to use it. As I will illustrate, although we began our work and presented our initial findings along these lines, such a paradigm became inadequate to explain the work I did in the classroom with Sarah and the students. As I reflected upon what I learned from the work, I realized that I needed a more flexible model, and that of narrative inquiry and storytelling works.

One of our strategies was to select teachers who understood, at least to some degree, the value of writing-to-learn. After examining syllabi from professors who used writing in their classrooms, we approached Sarah, whose syllabus for MGT 467, the capstone course for seniors majoring in Organizational Behavior, indicated that she asked students to do quite a bit of writing, both shorter and longer projects. She listed writing as one of many methods she used to "pursue the learning objective" ("Syllabus," Fall 1994, p. 1). We discovered that, in addition to her MBA and PhD in management, she also had a BA in English and was predisposed to see the value of writing.[10]

Sarah agreed to collaborate, and my colleague and I spent the remainder of the first semester as participant-observers. We observed Sarah's classes, interviewed the students and her, analyzed the students' written work, and examined all of the course materials. We worked to achieve triangulation of our data (Denzin, 1970). When the semester was over, we provided a recommendation report that was received enthusiastically and used as a basis for changes in her syllabus the following semester.

During the second semester, I continued to work as a participant-observer, noting what Sarah tried and how well it worked, but the dual roles of researcher and consultant caused some conflict. As a researcher, I believed that I should

[9] This is, in fact, what we did. We laid the foundations for a long-term project now known as the Howe Writing Initiative, which is overseen by the Howe Professor, an endowed position split between the SBA and the Department of English.

[10] Later in our collaboration, I realized just how differently she felt because she valued and used writing in her classrooms. She believed her use of writing put her at odds with other faculty in the SBA, saying once, "I often think I'm alone here in this world, Jim. Just think how much easier it would be if there were other people doing this. I heard the other day, someone said to me, 'No one else cares about this. Why do you?'"

maintain a distance, though as a consultant, I saw my role as one requiring more interaction. Other researchers have described similar conflicts. In her study of a 3-week writing project, Bonnie Sunstein (1994) described an incident in which she was "emerging from a shower, having a conversation about writing with a woman who was weeping because she was afraid to write" (p. 184). As she reflected upon and described the situation, she described how she had "a double persona here, like the mirror conversation in the shower room. . . . [She was] a writing teacher who [was] studying writing teachers and writing about it" (p. 184). She had to participate with the woman actively, intervening in the crisis; she couldn't maintain her distance. Nor later, when she had to write about it, could she maintain her distance. She was unable to "smother" her personal perspective.[11]

Some 2 years earlier, as a student and novice researcher in the English 302 class, I also had begun to see the need to eliminate the distance between researcher and subject(s). I knew that others had faced this issue; some have written about it. In her reflective piece on her dissertation work, Helen Dale (1996) described how, when faced with an ethical decision, she "felt [she] should remain aloof and let events play out" (p. 88), primarily because she was conducting "research." Her first obligation was not to the students or the class, but to the field.

My first clinical moment, one that colors everything else, occurred as a result of this conflict. In my section of MGT 467, there were 19 students, and we worked in two classrooms, splitting our time between a seminar room and the high-tech computer classroom. The seminar room had an oval seminar table with about 20 executive chairs. Around the exterior of the room, against the walls, were approximately six other chairs. Initially, I sat in one of these chairs and acted as a distanced observer, what many have called a "participant-observer." I anticipated having "as little effect as possible on the events and people under study" (Doheny-Farina, 1993, p. 255).

However, remaining a participant-observer while I was also serving an active assistant instructor who led several sessions in the computer classroom proved challenging. Thus, rather than remaining in the background, students saw me as a teacher, with expertise in writing and technology, which is how Sarah introduced me. I was a very visible "fly on the wall."

[11] Jennie Dautermann (1996) described a different, but related, situation when she outlined her position working in a hospital studying nurses and their communicative practices. She had to negotiate her status as both consultant and researcher, between working on research for her own dissertation and conducting work that would count as "billable hours." Such negotiation put her at odds with different groups (the hospital administrators and the nurses who had hired her), primarily because the "work and the observations occurred at the same time" (p. 245).

The fact that Sarah treated me as an expert made my work more pleasant and much easier.[12] I learned later that she had not always seen us as experts,[13] but by the time I learned this, we had been working together closely,[14] bound by many *ductile anchors*, a term I borrow from Whitman's (1931) poem "A Noiseless Patient Spider."[15] These anchors stemmed from some common beliefs about students, our obligations to them, and about writing and its place in the curriculum. Little by little, as Sarah and I worked and talked, sharing stories of our teaching and pedagogy, I could feel these ductile anchors take hold. As we began the second semester and talked more often about teaching, I knew that she valued me not only as someone who could help students use a listserv or write a paper, but also as a colleague and friend.

With the developing relationship between us, there was a subtle but pervasive shift in my stance. Because I had felt conflicted, I played the part of researcher during the first semester, limiting myself to observation, interviews, and analysis. However, as the second semester began, little by little, I found myself moving closer to the course material, the students, and the teacher. To put it in Van Manen's (1991) terms, I became interested. As my stance shifted, I felt as if I were in the midst of the class, and I felt obligated to participate, believing that if I interacted with the students, sharing my experience as a leader/manager, I might not only gain credibility, I could also get to know them better. I also believed that my experience might not only provide some interesting examples but would illustrate the kind of application of theory to practice that Sarah hoped to elicit from them.

[12]Such treatment is not always the case. Chiseri-Strater (1996) talked about how her "status as graduate student in composition marked [her] as powerless for some of the professors whose classrooms [she] studied" (p. 123).

[13]Initially she had accepted our offer because we were graduate students, but after our interviews, and more importantly, after we presented the recommendation report, her attitude changed.

[14]I did not have any expectations; I really thought I was mostly just helping you . . . . My courses had been praised, my assignments had received good feedback. As we got into our working relationship and I saw what you . . . could offer, I began to realize that the assignments and the learning could be more educational, more rewarding, and more constructive. This was a liberating opportunity for me because I knew intuitively before working with you that the assignments were ok but not great. Now I believe they are pretty challenging, interesting, and require more thoughtfulness on the part of students" (S. Michaels, personal communication, July 1997).

[15]In the poem, Whitman (1931) described "a noiseless patient spider" who launches "forth filament, filament, filament, out of itself" (p. 4) as a metaphor for the soul of the poet, working to establish "the bridge" with these "ductile anchor(s)." Although there are gulfs between the disciplines, there are many bridges that connect them, the most important one is the desire to teach students well.

This shifting stance and developing "interest" led me, after 3 weeks of class, to e-mail Sarah and ask her permission to share my thoughts and ideas about the course subject matter in addition to suggestions about the technology and its uses. When I asked, I was concerned about the move and what it might do to my status, but I trusted her advice not only because she had the authority as the teacher but also because I knew she had a background in qualitative research. She agreed, and when the opportunity arose, I began by occasionally volunteering comments in class and participating electronically. My decision to do so electronically had an ulterior motive. I was trying to help Sarah achieve a shift in the media students used to reflect upon course material, to encourage reflection via the medium of the class listserv, which at that time (fall 1994) was a relatively unused medium on campus. I wanted to see what would happen if one of our recommendations were enacted.

The discussion that occurred both on the list and in class was rich with stories. Students talked about relevant incidents and began to confront how they felt and what they might do if confronted by a problem in an organization.[16] I felt a change in my status as well. Students started talking directly to me in and out of class. And within 2 days, I received an invitation from Sarah, via e-mail, that explained she had been talking with some of the students in the class who had suggested "it would be nice to have [me] sit at the table with [them]" (S. Michaels, personal communication, February 28, 1995).

Her invitation was what I would call my first "clinical moment." By intervening and participating in the readings and discussions (even if I limited that participation), I was able to overcome the objection North (1987) described in his criticism of Odell and Goswami's (1986) research: "Odell and Goswami might be said, via their interviews, to get a glimpse of the workers' imaginative universe, but from an Ethnographic point of view they have no way to make sense of what they see" (p. 302). By joining the class at the table, I was able to get a glimpse from the students' perspective, albeit one colored by many other sets of lenses. Yet I believe that my seat helped me to provide a better understanding of the students' perspectives. Sitting at the table (or later around the circle) effectively rendered me more of a collaborator (with both students and teacher) than teacher. What I learned from observing began to be funneled back into the classroom. I became more of a *dialogic partner* with the students and teacher than a researcher.

[16] In his book, *The Tact of Teaching,* Max van Manen (1991) described an important stance that teachers take: they "stand in the midst (inter esse), the roots of the word "interest. . . . To be interested in something is to stand in the midst (inter esse) of something, to take part in, to maintain a caring relation to something. To be interested is to be intensely present to something or somebody" (p. 196).

## Clinical Moment #2

When I was growing up, as part of the discussion surrounding President Kennedy's assassination, my parents always told me that I'd remember certain moments in history and where I was at the time. Their words remained true as I can recall moments such as President Nixon's resignation, the Space Shuttle Challenger's disintegration minutes after launch, and 9/11, which becomes the catalyst for this clinical moment. On the absolutely gorgeous fall day of September 11, 2001, I was teaching a section of English 3774, Business Writing, that was to meet at 9:30 a.m. The plan that day was to outline our semester projects using the proposal genre. Our partners were the YMCA at Virginia Tech (the Y) and the town of Blacksburg. By the time students came to class, many had heard about the incident, and even before class began, we turned on the TV and went online. My second moment is actually more than a moment as it involves a subtle but pervasive shift in the class' desire to act for some common good. When I first wrote about service learning as an act of virtue (Dubinsky, 2002b), I chose not to focus on the connection to 9/11 for many reasons, some of them personal. But without a doubt, the discussions we held that term about the role of higher education, the importance of service to one's community/country, and the work they were doing in this class, I know with certainty that the students' convictions, even though not explicitly linked, were connected to that terrible morning when planes became missiles and our country and the world were forever changed.

For the 3 years prior, I had been the only tenure-track faculty member with a background in professional communication. My mission had been to design a program in Professional Writing for English majors and make our service courses (Business and Technical Writing) more responsive to the students and departments sending those students to us (Dubinsky, 2010b). As a result, although I had spent some time volunteering and working with organizations such as the Y, I had spent the majority of my time and energy in program and course design/redesign. For many reasons, not the least of which was my success in finding colleagues with whom I would partner, all of whom were instructors at the time, we had been able to convince the dean to hire two additional tenure-track faculty. With their presence, I now had a bit more time to devote to my community partners, and the tragedy of 9/11 heightened that desire to provide more assistance. And as a result, many instructors teaching our service courses, as well as some of the faculty teaching the courses in the program we were building, began focusing on the advantages of experiential learning.

This second "clinical moment" involves a shift in my role as partner that led to a deeper and longer-term commitment not only on my part but also on the part of many faculty in our program. In my particular class that semester, our partner, the Y, needed help with their programs, which included international students and their families (Dubinsky, 2008). In addition, we were seeking funds for a before- and after-school program that became known as the Y-Zone, which would be held at one of the elementary schools. During this process, we, and I in particular, had to learn a lot about the Y's financial needs. And because of our work, the Y's executive director asked me to join the board, a board I sat on for 10 years, to include a term as chair. It was that moment, the moment of being asked to commit to something quite a bit beyond and outside the service learning partnership and occasional volunteer project, that had a powerful impact on my teaching and the overall program design not only for the undergraduate program but also the PhD program in Rhetoric and Writing instituted in 2007. Equally important (but not as relevant to this chapter), it was formative in the work I was asked to do between 2007 and 2011 as chair of the Virginia Tech Task Force on Student Engagement and then as the founding director of the Center for Student Engagement and Community Partnerships (Dubinsky, 2010b).

The work I dedicated myself and students to, as well as provided leadership to the university about, has been described as "closing the gap between the world of ideas and the theoretical reflexes that animate faculty culture, on the one hand, and our students' preference for concrete applications of knowledge and for active learning, on the other" (Cooper, as cited in Barker & Brown, 2009, pp. 22–23). This work had a direct impact on my efforts toward program building, work that I describe as based on reverence and what Paul Woodruff called "the paradox of respect" (as cited in Dubinsky, 2010a, p. 278). Woodruff explained that reverence is "the capacity for certain feelings where they are due" and one of those is respect, which involves "a felt recognition of a connection growing out of common practices" (as cited in Dubinsky, 2010a, p. 278).

## The Moments and Their "Meaning"

By choosing the first moment, I'm highlighting dilemmas I faced due to competing demands caused by the multiple roles I played as researcher, collaborator, consultant, and teacher (Brueggemann, 1996; Chiseri-Strater, 1996; Dale, 1996). I am raising the issue of the teacher-researcher's stance and what is implied when the barrier or distance between subject and researcher is eliminated. I'm outlining the "dilemma of whether merely to record or to change classroom behavior, and the level at which to direct this change is especially significant for someone who is a full participant in the environment under study" (Ray, 1996, p. 292). Most important (at least to the purpose/aim of this volume), I am arguing that program administrators must always be both participants and researchers who, by nature of their commitment to students and to the teaching/learning process, are involved. The second story about my partnership with the YMCA at VT documents the centrality and significance of building a mutually reciprocal relationship as a strategy to create an effective learning environment and an effective professional writing program.

By telling these stories, I offer an alternative to the standard "realist tales" that are often expected of researchers, but I do more, I believe, than produce a "wild and wooly convoluted tract . . . that suck[s] its author (and reader) into a black hole of introspection" (Van Maanen, 1988, p. 92). Instead, I am "an interpreter of indigenous texts" (p. 75) and a reflective practitioner who uses what he learns from that interpretation to create a procedural knowledge of "how to" create change in the classroom and thus work toward the kinds of goals some scholars have labeled emancipatory (Freire, 1993). These goals are part and parcel of the ethical dimension of *phronesis* that Aristotle attributed to historical figures such as Pericles. They are the result of a "reasoned state (*hexis meta logou*)" (Dunne, 1993, p. 264) that must be "done [*prakton*] . . . for good action [*eupraxia*]" (Aristotle, *E.N.* 1139a36-1139b4). The *technites*, in this case a teacher, partner, or administrator, subordinates her concerns to those wider concerns of students, colleagues, or community partners to accrue excellence (*aretê*). As Dunne (1993) argued convincingly,

> If medicine enables one to produce health *and* disease, then one might say that it ceases to be a mere *dunamis* (*tôn enantiôn*) and accrues excellence only in a doctor reliably committed to *health*. The technical competence of such a doctor will be infused with a solicitude for his patients. (p. 265, emphasis in original)

This solicitude is evident in Coles' (1989) stories about patients, a solicitude inculcated by a desire to learn by using a narrative process. He succeeded because he used the narrative method and exhibited an attention and care for the health of his patients. This is caring and solicitude that teachers, community partners, and administrators would do well to exhibit.

In addition, as I tell these stories, I believe I offer something useful to our field in terms of what I learned about collaborating. Embedded in this discussion is information I have learned about how teachers, who are also partners and administrators, "articulat[e] the ways of knowing that are central to particular domains" (Langer, 1992, p. 70). I believe my discussion also emphasizes the importance of process: having a *researcher collaborate with a teacher or a partner* not only enables a researcher to learn more about reflective practice, it also empowers the teacher/colleague as a reflective practitioner and co-researcher (Grego, 1994, p. 226).

## Academic Prose, Boundaries, and "Restorying"

The more often I read Doheny-Farina's (1993) "Research as Rhetoric" and Robert Coles' (1989) *The Call to Stories,* the more I realize that what I have to say couldn't and shouldn't be said in the language of the "academy" and its researchers. Coles felt conflict about whether he was being scientific and professional when he worked with stories as opposed to problems, when he related narratives instead of clinical analyses.[17] Yet despite (or perhaps as a result of it), he chose to tell the story of his own "apprenticeship" as an intern in a hospital in a book about valuing those stories and that process.[18] I too am using stories as a way of knowing and a way of sharing. Through collaboration and storytelling, I am able to describe what I learned through my various apprenticeships and how that knowledge informed my practice, creating traces that are etched into the program design I helped create.

As Coles (1989) did, I realize that my work has been work *with* people, not *on* people (Fleischer, 1995). During the first collaboration in graduate school, I began as an apprentice, much in the same position as Coles was in the hospital. Both my business school colleague and I had the education and backgrounds to be considered "expert," yet I for one felt somewhat unsure of myself among the wealth of theory and "formulations" required by those senior experts in our discipline. A similar situation existed when I began working as a partner who was expected to, through the process of service learning, help build the capacity of an established nonprofit organization. And in considering my role as program designer and administrator, I was also a novice, depending on building relationships and creating a sense of trustworthiness within a department filled with literature colleagues. Choosing to talk about these instances now, to "restory" them, allows me to "reexamine and reinterpret" as I create what Clandinin (1996) called "professional knowledge landscapes" (p. 28). It also helps to re-create the events in my mind, events that are examples for me of what Myerhoff and Metzger called "great moment[s] of teachability" (as cited in Sunstein, 1996, p. 178).

In addition, if, as Doheny-Farina (1989) and others have speculated, what an ethnographer writes is not research but instead a "realistic fiction" (p. 253), then it may be more important for me to tell my story more as a "confessional tale" (Van Maanen, 1988) than a realist one and let its authority derive from my

[17]Sarah and I have reported on the content of these stories and how, we believe, they led to improved written analysis, not only of the issue of diversity but other issues in the course as well.

[18]He actually said that he was "a fledgling, an apprentice in psychiatry, who wasn't at all sure he *was* the doctor and who hadn't yet become as time-conscious as one becomes when in practice" (Coles, 1989, p. 4, emphasis in original).

attempt to be rhetorical.[19] Actually, although I understand why Van Maanen (1998) chose the label "confessional tale," the term strikes me as de-meaning.[20] I would prefer to change the term, shifting away from Van Maanen's taxonomy, useful as it is, to *tales of theory in practice*, borrowing from Coles (1989) and Berthoff (1987). These tales, "generated in dialogue among teachers" (Berthoff, 1987, p. 29), represent a valid form of research, part of the larger scholarship of teaching (SoTL) effort that began when Boyer (1990) proclaimed a need for the academy to recognize other forms of scholarship.

Perhaps the most important point to highlight in these stories is one I made earlier: because the teacher or program administrator is bound up in "a network of other people," and one of the goals is to move those people as far along the road to health (or in this case, knowledge or understanding the role of writing plays in constructing knowledge), it seems to me that it is the teacher's *duty* or *obligation* to intervene. Aristotle's discussion of the notion of "good action," when combined with a master craftsperson (*technites*) who is not shaping an impersonal form such as wood or clay but is working with agents, helps us understand the complex relationship of *techne* and *phronesis*—of art and ethics. As teachers, our "reasoned capacity to make" (*techne*) must include reflection not only about the craft but also about the persons with whom we are "bound up." The means and ends are intimately interwoven.

There were tangible benefits from both collaborations. As we worked together, reflecting on the problem, on our work together, and on the students' papers, more of what Chris Argyris (1994) called "double-loop learning" occurred (p. 78). The students became more involved in discussions of their rhetorical strategies as they wrote, interpreted, and negotiated texts. We all learned to see evaluation from different perspectives, and we learned why teachers should write standards that students can understand. Thus, our acts of assessment and the resulting discussions with the students became "episodes of learning" about the subject, writing, and teaching—especially teaching. In both cases, the two partners were forced to reflect more frequently about the social and interactive

[19]Rhetorical narrative inquiry has educative possibilities. Clandinin (1996), for instance, saw potential for teacher education. The stories that get told benefit the student teachers, the cooperating teachers, and a larger audience. Paying attention to that larger audience is essential: "Something of the spirit of action of the participants, students, teachers and university teachers needs to find a place in these stories. As in the accounts of research, issues of representation, and audience are central concerns" (p. 135).

[20]I hyphenate this word to emphasize the taking away of value and meaning, to highlight the often perceived lower status. Sunstein (1994) made it clear that she didn't want to write a "confessional tale" because, in 1994 when she wrote it, such a tale wasn't considered sufficient for the gatekeepers (pp. 28–29). Later, when she reflected upon that narrative (1996), there is much that might be called "confessional." From my background of Catholicism, the term calls to mind revealing indiscretions, failures, and foibles.

aspects of teaching and learning (Richert, 1992, p. 187). At the same time, the "episodes of learning" didn't end. Sarah and I continued to talk about the issue of evaluation and ways for her to help students learn the language of their discourse communities. And I continue to work with the YMCA to help them "build community through learning, leadership, and service."

## CONCLUSION

In this chapter, I have told stories of collaboration and highlighted the critical role of relationship building. In each collaboration, I wanted "it all," as Dale (1996) put it. I wanted to be a participant-observer, colleague, friend, teacher, and researcher. I expected to learn a good deal about teaching, dialogic partnerships, procedural knowledge, and what Gibbons (2001) called "knowledge in action" (pp. 12–13). My primary goal in every instance has been to improve the learning of students in classrooms (Ray, 1996). In every instance, my colleagues, partners, and I made honest efforts to work "for the good of the governed." To achieve that good, we had to reflect upon our work, with a goal of "refining the skills [and focusing] on the *why* of what [we] were doing" (Dunne, 1993, p. 333).

Our relationships were similar in nature to one Aristotle emphasized in his discussion of *techne* and *technites*. In *De Anima* he said, "[the *technites* must state the reasons why he struck his blow in such a way to effect this" (1.1.641a13–14, as cited in Dunne, 1993, p. 332). He must state those reasons to someone else, and that someone else must be another craftsperson, aware of the situation and the material. In our relationship we constantly asked each other to "state the reasons why."

In this work, I have been fortunate to research, teach, and work toward program design and administration with colleagues and partners whose obligations were directed toward the classroom. My work helped me to *see* what is really significant in different situations (Dunne, 1993, p. 308), and in representing that work, I have come to realize that sharing stories is a viable means of producing knowledge. Most importantly, I have come to realize that all prospective teachers, nay all teachers, need practice and opportunities to collaborate, research, and build with others, individuals with whom they can face "clinical"/critical moments of "teachability."

## REFERENCES

Argyris, C. (1994, July/August). Good communication that blocks learning. *Harvard Business Review, 72*(4), 196–204.

Aristotle. (1941). De anima (J. A. Smith, Trans.). In R. McKeon (Ed.), *The basic works of Aristotle* (pp. 553–603). New York, NY: Random House.

Aristotle. (1941). Politics (B. Jowett, Trans.). In R. McKeon (Ed.), *The basic works of Aristotle* (pp. 1113–1324). New York, NY: Random House.

Aristotle. (1954). *The rhetoric & the poetics of Aristotle* (W. R. Roberts, Trans.). New York, NY: Modern Library.

Aristotle. (1962). *Nicomachean ethics* (M. Ostwald, Trans.). Indianapolis, IN: Bobbs-Merrill.

Barker, D. W. M, & Brown, D. W. (Eds.). (2009). *A different kind of politics.* Dayton, OH: Kettering Foundation.

Berthoff, A. E. (1987a). From dialogue to dialectic to dialogue. In D. Goswami & P. Stillman (Eds.), *Reclaiming the classroom: Teacher researcher as an agent of change* (pp. 75–86). Upper Montclair, NJ: Boynton/Cook.

Berthoff, A. E. (1987b). The teacher as researcher. In D. Goswami & P. Stillman (Eds.), *Reclaiming the classroom: Teacher research as an agency for change* (pp. 28–38). Upper Montclair, NJ: Boynton/Cook.

Boyer, E. (1990). *Scholarship reconsidered: Priorities of the professoriate.* Washington, DC: The Carnegie Foundation for the Advancement of Teaching.

Britzman, D. P. (2003). *Practice makes practice: A clinical study of learning to teach.* Albany, NY: SUNY Press.

Brueggemann, B. J. (1996). Still-life: Representations and silences in the participant-observer role. In P. Mortensen & G. Kirsch (Eds.), *Ethics and representation in qualitative studies of literacy* (pp. 17–39). Urbana, IL: National Council of Teachers of English.

Chiseri-Strater, E. (1991). *Academic literacies: The public and private discourse of university students.* Portsmouth, NH: Boynton/Cook.

Chiseri-Strater, E. (1996). Turning in upon ourselves: Positionality, subjectivity and reflexivity. In P. Mortensen & G. Kirsch (Eds.), *Ethics and representation in qualitative studies* (pp. 115–133). Urbana, IL: National Council of Teachers of English.

Clandinin, D. J. (1986). *Classroom practice: Teacher images in action.* London, UK: Falmer.

Clandinin, D. J. (1993). *Learning to teach, teaching to learn: Stories of collaboration in teacher education.* New York, NY: Teachers College Press.

Clandinin, D. J. (1995). Still learning to teach. In T. Russell & F. Korthagen (Eds.), *Teachers who teach teachers* (pp. 25–31). London, UK; Washington, DC: Falmer.

Clandinin, D. J. (1996). Teachers' professional knowledge landscapes: Teacher stories—stories of teachers—school stories—stories of school. *Educational Researcher, 25*(3), 24–30.

Clandinin, D. J., & Connelly, F. M. (2004). *Narrative inquiry: Experience and story in qualitative research.* San Francisco, CA: Jossey-Bass.

Coles, R. (1989). *The call of stories.* Boston, MA: Houghton Mifflin.

Dale, H. (1996). Dilemmas of fidelity: Qualitative research in the classroom. In P. Mortensen & G. E. Kirsch (Eds.), *Ethics and representation in qualitative studies of literacy.* Urbana, IL: National Council of Teachers of English.

Dautermann, J. (1996). Social and institutional power relationships in studies of workplace writing. In H. Mortensen & G. Kirsch (Eds.), *Ethics and representation in qualitative studies of literacy* (pp. 241–259). Urbana, IL: National Council of Teachers of English.

Denzin, N. K. (1970). *The research act.* Chicago, IL: Aldine.

Doheny-Farina, S. (1989). A case study of one adult writing in academic and nonacademic discourse communities. In C. Matalene (Ed.), *Worlds of writing: Teaching and learning in discourse communities of work* (pp. 17–42). New York, NY: Random House.

Doheny-Farina, S. (1993). Research as rhetoric: Confronting the methodological problems of research on writing in nonacademic settings. In R. Spilka (Ed.), *Writing in the workplace: New research perspectives* (pp. 253–267). Carbondale, IL: Southern Illinois University Press.

Dubinsky, J. M. (1998). *Learning the Möbius loop of theory and practice: Reflections on the têchne of teaching writing.* Dissertation. Miami University, Oxford, OH.

Dubinsky, J. M. (2002a). More than a knack: Technê and teaching technical writing. *Technical Communication Quarterly, 11*(2), 131–147.

Dubinsky, J. M. (2002b). Service-learning as a path to virtue: The ideal orator in professional communication. *Michigan Journal of Community Service Learning, 8*(2), 61–74.

Dubinsky, J. M. (2008). When the global is local: Building a local, global community through partnership and pedagogy. In D. Starke-Meyerring & M. Wilson (Eds.), *Designing globally networked learning environments: Visionary partnerships, policies, and pedagogies* (pp. 170–184). Rotterdam, The Netherlands: Sense.

Dubinsky, J. M. (2010a). A *techne* for citizens in design discourse. In D. Franke, A. Reid, & A. Di Renzo (Eds.), *Design discourse: Composing and revising programs in professional and technical writing.* West Layfayette, IN: Parlor.

Dubinsky, J. (2010b). Program review: The land-grant way—Connected knowing and the call of service. *Present Tense: A Journal of Rhetoric in Society, 1*(1), 1-9.

Dunne, J. (1993). *Back to the rough ground:* Phronesis *and* techne *in modern philosophy and in Aristotle.* South Bend, IN: University of Notre Dame Press.

Eisner, E. (1988). The primacy of experience and the politics of method. *Educational Researcher, 20,* 15–20.

Fleischer, C. (1995). *Composing teacher-research: A prosaic history.* Albany, NY: SUNY Press.

Freire, P. (1993). *Pedagogy of the oppressed* (M. B. Ramos, Trans.). New York, NY: Continuum.

Gates, R. (1990). Understanding writing as an art: Classical rhetoric and the corporate context. *Technical Writing Teacher, 17,* 50–60.

Gee, J. P. (1985). The narrativization of experience in the oral style. *Boston University Journal of Education, 167,* 9-35.

Gelinas, U. J., Rama, D. V., Jr., & Skelton, T. M. (1997). Selection of technical communication concepts for integration into an accounting information systems course: A WAC case study. *Technical Communication Quarterly, 6*(4), 381–401.

Gibbons, J. (2001). Knowledge in action. *Philosophy and Phenomenological Research, 62,* 579–600.

Grego, R. C. (1994). Writing academic autobiographies. Finding a common language across the curriculum. In P. Sullivan & D. Qualley (Eds.), *Pedagogy in the age of politics* (pp. 214–229). Urbana, IL: National Council of Teachers of English.

Heilbrun, C. (1988). *Writing a woman's life.* New York, NY: Norton.

Hirschorn, L. (1991). Organizing feelings toward authority. In D. Schon (Ed.), *The reflective turn: Case studies in and on educational practice* (pp. 111–126). New York, NY: Teachers College Press.

Knoblauch, C. H., & Brannon, L. (1988). Knowing our knowledge: A phenomenological basis for teacher research. In L. Smith (Ed.), *Audits of meaning: A festscrift in honor of Ann E. Bertoff* (pp. 17-28). Portsmouth, NH: Boynton/Cook.

Langer, J. A. (1992). Speaking of knowing: Conceptions of understanding in academic disciplines. In A. Herrington & C. Moran (Eds.), *Writing, teaching and learning in the disciplines* (pp. 69–85). New York, NY: Modern Language Association of America.

Mishler, E. G. (1979). Meaning in context: Is there any other kind? *Harvard Educational Review, 49*, 1–19.

Mishler, E. G. (1986). *Research interviewing: Context and narrative.* Cambridge, MA: Harvard University Press.

Mishler, E. G. (1990). Validation in inquiry-guided research. *Harvard Educational Review, 60*, 415–443.

North, S. (1987). *The making of knowledge in composition: Portrait of an emerging field.* Portsmouth, NH: Boynton/Cook.

Odell, L. (1992). Context-specific ways of knowing. In A. Herrington & C. Moran (Eds.), *Writing, teaching, and learning in the disciplines* (pp. 86–98). New York, NY: Modern Language Association of America.

Odell, L., & Goswami, D. (Eds.). (1986). *Writing in nonacademic settings*. New York, NY: Guilford.

Perkins, D. (1986). *Knowledge as design*. Hillsdale, NJ: Lawrence Erlbaum.

Ray, R. (1992). Composition research from the teacher-research point of view. In G. Kirsch & P. A. Sullivan (Eds.), *Methods and methodology in composition research* (pp. 172–189). Carbondale, IL: Southern Illinois University Press.

Ray, R. (1993). *The practice of theory.* Urbana, IL: National Council of Teachers of English.

Ray, R. (1996). Afterword: Ethics and representation in teacher-research. In P. Mortensen & G. E. Kirsch (Eds.), *Ethics and representation in qualitative studies of literacy* (pp. 287–300). Urbana, IL: National Council of Teachers of English.

Richert, A. E. (1992). The content of student teachers' reflections within different structures for facilitating the reflective process. In T. Russell & H. Munby (Eds.), *Teachers and teaching* (pp. 171–191). London, UK: Falmer.

Saville-Troika, M. (1989). *The ethnography of communication* (2nd ed.). London, UK: Basil Blackwell.

Schön, D. A. (1983). *The reflective practitioner: How professionals think in action.* New York, NY: Basic.

Sullivan, P. A., & Qualley, D. J. (Eds.). (1994). *Pedagogy in the age of politics.* Urbana, IL: National Council of Teachers of English.

Sunstein, B. S. (1994). *Composing a culture*. Portsmouth, NH: Boynton/Cook.

Sunstein, B. S. (1996). Culture on the page: Experience, rhetoric, and aesthetics in ethnographic writing. In G. Kirsch, & P. Mortensen (Eds.), *Ethics and representation in qualitative studies of literacy* (pp. 177–201). Urbana, IL: National Council of Teachers of English.

Tebeaux, E. (1988). The trouble with employees' writing may be freshman composition. *Teaching English in Two Year Colleges, 15,* 9–19.

Van Maanen, J. (1988). *Tales of the field: On writing ethnography.* Chicago, IL: University of Chicago Press.

Van Maanen, M. (1991). *The tact of teaching.* Albany, NY: SUNY Press.

Whitman, W. (1931). A noiseless patient spider. *Leaves of grass.* New York, NY: Aventine.

CHAPTER 6

# Leaders Becoming Transformed

***Meg Morgan***

Becoming an effective transformational leader is like becoming an effective parent; the early stages of both are marked by much attention to creating or changing behaviors; the later stages, I would contend as a parent who now has two daughters in their 30s, is much more like collaboration and sharing. Both processes are just that—processes. The later stages are easier if the early stages contain clear expectations about values and behaviors. I suppose one could be a transformational leader right from the beginning of a project, childrearing or any other, but I would hate to have to collaborate and share with a 2-year-old. This chapter looks at two different styles of leadership, what Sashkin and Sashkin (2003) called transactional and transformative. They are marked in some ways by development—like raising a child—but are more clearly marked by characteristics and attitudes toward the followers by the leader. Transactional leaders focus on getting the job done; transformational leaders focus on transforming others into leaders to get the job done.

Leadership exists within two enactments of power: personal power and prosocial power (Sashkin & Sashkin, 2003, pp. 58–59). Leaders who work within the construct of personal power usually work for personal gain. We've met these leaders: "They exercise power in order to differentiate themselves from everyone else and prove their special status." People in this power game usually benefit in ways that do not necessarily help anyone but themselves. They get raises, promotions, and special handouts. They work for personal visibility. The other power type is prosocial power. Those who work within the prosocial power construct work to "attain organizational goals. To achieve such goals, they typically share power; not only [do] they try to influence others, they [are] themselves open to being influenced" (p. 58). Transactional leaders who are personally driven often work alone—they don't empower others because, in doing so, they perhaps would not achieve what they desire. However, one can be a transactional leader and also prosocial, the goal of which would be to work with others so everyone would have control. Transformational leaders can also be motivated by either personal power or by prosocial power. Those motivated by personal power try to inculcate their own values into their followers; those motivated by prosocial power try to identify shared values. Transformational leaders attempt to influence others to become transformational leaders by articulating values and beliefs in such a way that others adopt those values and beliefs as their own. This is not manipulative. In fact,

> the transformational leader's work is to first derive a set of values and beliefs of the current community of organization members (including the leader). Transformational leaders then act to make these values and beliefs explicit so that they can be shared consciously. (p. 72)

Thus, the major relationship in transformational leadership is between the leader and the followers, and the major goal is to make the followers independent thinkers and leaders. So transformational leadership is a social endeavor in its transformative process.

Given this context (personal power and prosocial power within which one could be a transactional leader or a transformational leader), let me give you a short history of the development of leadership within the Technical/Professional Writing program at the University of North Carolina, Charlotte (UNC Charlotte). The development seems to follow a prosocial power structure that moves over almost 20 years from a transactional to a transformational paradigm.

## LEADERSHIP TRANSFORMATION AT UNC CHARLOTTE

At UNC Charlotte, we have a very successful program in Technical/Professional Writing. Like most programs, its success is based on the teaching and research competence of its faculty and the fact that the field, even in economic hard times, is still viable. Students in other majors see the value of taking our courses to enhance their own majors, so we must restrict access so there will be room in courses for those enrolled in the program. Yet the success of the program is also based on the efforts we have made to embed our program into the activities of the university community as well as the community at-large. I have directed the program in two stages: in its very beginning years, from 1988 to 1994 and now in its later more mature years, from 2003 to the present. I am a leader with "prosocial power." I have always looked into the community to work with people who share my values about the program. If necessary, I have worked to shape my values so we can all have some common ground. One way to look at the growth and success of our program is to tie it to the idea of growth and change in leadership styles, from a leadership style that was prosocial/transactional to one that is now prosocial/transformational. And although I was the initial coordinator of the program and coordinate it currently, I did not always direct it. In every instance, however, the other directors of the program shared common values and assumed leadership roles that have moved the program forward.

My initial goal as its director was to involve other people in the process of moving our program from a 1-course offering to a program that contained an undergraduate minor and a master's in English with a technical/professional writing emphasis. I knew I could not do this alone (nor did I want to), so I invited others into the process. And prosocial power/transactional leadership was exactly what our new program needed. In the late 1980s, few people had heard about, understood, or cared about technical communication. To many, technical writing was "skill-based" writing, formulaic in the worst sense of the word. According to many, students in the program did not have to be very creative or even smart, they just had to follow the formulas and their writing would be good. Much of the leadership of the first few years involved teaching people that technical writing was creative, that doing it successfully required rhetorical skills, and that the writers had to manipulate not only advanced rhetorical skills but also technical skills. One way I accomplished this was to "empower others to achieve goals," and thus, according to the prosocial development grid, I enacted a prosocial form of transactional leadership through hard work, a clear vision of the discipline, a sense of where we wanted to go, and prosocial persuasion (this would be good for the whole English Department). Because we worked to get others to buy into our values, we were transactional leaders.

At about the same time, we began our collaboration with the local chapter of the Society for Technical Communication (STC). We created an advisory board made up of local writers and leaders in the technical writing profession, most notably the Metrolina Chapter of the STC. We consulted with them for several years, a collaboration that involved, most notably, the initiation of a university-based Technical/Professional Writing Exhibit of Student Work. The STC chapter allowed readers to judge the students' work, it provided food for the exhibit, and awarded prizes for the students whose exhibits won in their categories. For several years, recruiters from the area came to talk with students about jobs in the profession. Over the years, program directors sent out invitations, provided the space and, obviously, the student work. Several deans from the university attended the exhibit. This exhibit is still held every year and is still the highlight of the spring semester for many students—they know it is coming and work toward it. Faculty motivate students to write for the exhibit; faculty help organize the event; and one faculty member for years has had his students move tables from classrooms to the open space in the atrium where the event is held. This collaboration with the STC chapter is a good example of prosocial transactional leadership: a person (the program director) working with the university and local community to get the job of creating and defining a technical/professional writing program done. Other examples of this prosocial/transactional style of leadership might include bringing STC members to campus for one of its monthly meetings, getting students involved in community-based editing projects, and working with the graduate director in English to recruit graduate students.

Over the years, my leadership style has changed. Although early in the program, I was a proactive/transactional leader, now I am what might be called a prosocial/transformational leader. The transformative leadership is mostly connected to my work with students in that I have helped to transform students from members of a university to members of a professional working community. Perhaps the most notable example of this is the transforming of students into active and caring members of their professional community. When I teach my technical/professional writing courses (and this is true with other faculty in the program, so we really are sharing values), much work includes working with the community on projects: students in Dr. Deborah Bosley's grant-writing course wrote a grant that brought money into a local community-based organization; one student designed documents for a writing center in Kenya. Our programmatic values are very much paying back to the communities in which we learn and live. And many students seem to have learned this lesson after they graduate. Several members of the STC board are former students. They bring to the board the values that the program has helped them form: collaboration, creativity, hard work, and giving back. We still put on the exhibit every year, but it has become more faculty-driven rather than director-driven.

In each example, the true transformational leader has been the UNC Charlotte community of technical communication scholars and teachers who have worked over the years to transform students into community leaders with whom we share values.

## CREATING TRANSFORMATIONAL LEADERS

I would guess that many of us in the teaching field are prosocial; it would be difficult to teach and constantly interact if we did not see our work as benefitting society. I would guess that many of us running programs in universities and colleges are transactional leaders. I would not guess that many of us are transformational leaders in that we consciously and deliberately try to transform our "followers" by articulating our shared values and motivating them to become transformational leaders. Program directors must do more than direct or manage a program, even if that direction is democratic. We certainly have the responsibility to, in the most extreme sense, create new people, teaching them not just subject matter but also human and social values in the ways they work with others.

Assuming a prosocial power, here are some ideas about ways to enable transformational leadership skills and at the same time help transform faculty and students into strong transformational leaders.

- Examine with your faculty and students the values embedded in the community—both the university community and the local community. This examination and ultimate understanding of the community can enable them to be more effective and valuable members of those communities.

Certainly this examination can take place in a classroom, but it could also take place at a local professional organization meeting, at on-campus meetings, and in student chapters of the STC.

- Empower students with real responsibility. This was done in our program by having them write real grants and design real websites. I have asked my editing students to edit documents for members of the English faculty. By taking this risk, I am empowered, but they are too. Initially transactional because it is part of a class assignment, the work transforms students as they assume responsibilities and take on opportunities for paid work. In fact, after taking the editing class, several students have edited documents for the English department and for individual faculty.
- Create relationships with members of the community wherein all members in the relationships share the same vision and are equally empowered. Our relationship with our local STC has evolved over the years as students have matured in the community. One interesting change last year was that the initial meeting of the STC occurred at a community park where we shared drinks, hamburgers, and hot dogs. This risk brought in a few new members. Next year, should this same meeting recur, I sense that more new people will come—the value that social interactions are as important as work-based ones will be reinforced.

One of the most outstanding examples of prosocial transformational leadership occurred in the creation of the Carolinas affiliate of the Writing Program Administration (WPA). About 10 years ago, I was a member of the executive board of the national WPA and learned about their efforts to establish affiliates at the state level. With the help of the Council and the enthusiasm of many WPAs in North and South Carolina, we set up the affiliate—a truly prosocial, transformational leadership-based organization. Members have met for many years and have come to an understanding of common and shared values—professional, theoretical, and social. We share leadership of the organization in a way that reflects a prosocial/transformative model; and one of our goals is to transform others and ourselves. Members of the technical communication community attend these meetings and share in the collegiality that transforms its members.

## CONCLUSION

Many of us do not begin as transformational leaders—perhaps we are transactional because that's the way we are. However, if change is to be accomplished and accepted by all members of the community, leadership must be transformational. Establishing shared values and working toward shared goals is what transformational leadership is all about. And one of the ultimate outcomes of such a philosophy is personal transformation. Transformational leaders transform others and also become transformed. Like when we raise children, we are changed forever.

## REFERENCE

Sashkin, M., & Sashkin, M. G. (2003). *Leadership that matters: The critical factors for making a difference in people's lives and organizations' success.* San Francisco, CA: Berrett-Koehler.

CHAPTER 7

# The Challenges of Offering a Technical Writing Program in a 2-Year College

***Ritu Raju***

In the 3 decades following the establishment of the Council for Programs in Technical and Scientific Communication (CPTSC), programs in technical communication have grown in number and strength. However, much of this growth has been restricted to the arena of the 4-year college or university. In his 2008 keynote address at the International Professional Communication Conference (IPCC) at Montreal, Saul Carliner suggested that community colleges or 2-year colleges are excellent venues for technical communication programs. Yet few community colleges offer anything beyond the service course for technical writing. A quick survey of Society for Technical Communication's (STC) academic database shows the following numbers:

- Programs offering associate of arts or science in the United States: 7
- Programs offering an undergraduate diploma: 0
- Programs offering an undergraduate certificate: 24. Of these, only two are community colleges; the others are all universities.

In contrast, the number of universities offering bachelors' programs in the field was 68.

These numbers reflect a paradox that forms the basis of my experience and my chapter. On the one hand, community colleges seem to be the perfect venue for introductory programs in technical communication that will provide industry with entry-level technical writers; on the other, specific problems, institutional roadblocks, and perceptual fallacies prevent community colleges from offering such programs. As early as 1979, Brockman suggested that "the problematic consequences of their unique design features" prevented community colleges from achieving success in sustaining technical communication programs. Although early technical communication programs in the 1970s saw encouraging growth, increasing from three programs in 1972 to six by 1976 (Brockman, 1979), the pace of growth slowed down by the end of the decade. The factors that resulted in this decline seem to hold true even today; they continue to shape and inform the nature and course of technical communication programs in 2-year colleges. These factors are composed of the following characteristics:

- A close working relationship with industry
- A 2-year duration
- Open admissions
- Newness

In this chapter, I discuss some of these issues in light of my experiences in setting up a technical communication program at a 2-year institution. I must clarify that this program is in the process of creation and reinvention; it seeks to replace a previous program that has since been discontinued. Thus, although I report the findings of our needs-assessment efforts and preliminary research findings, much of the program development is still underway. I will address Brockman's (1979) factors in the light of our own experience as well as discuss the results and findings of our program-assessment activities. The latter fall into the following categories:

- Focus group findings
- Transfer of credits/integrating with bachelor's programs
- Assessing needs of service area

Before I discuss our experiences, it would be appropriate to offer some background information about the institution and its experience with previous technical writing programs.

## SOME HISTORICAL BACKGROUND

Premier Community College comprises six colleges with more than 15 campuses. The college offers both academic and workforce programs. Although the academic programs culminate in associate's degrees, the workforce programs lead to degrees, certificates, or Marketable Skills Awards (short certificates geared toward quick entry into entry-level jobs). One of the colleges had been offering several programs in technical communication, including variations of all these: an associates, some certificate programs, and a couple of Marketable

Skills Awards (MSA). Over a period of time, the department's focus shifted to digital communication for three main reasons: a shift in the area's job needs, lack of sustained marketing and growth initiatives, and changing priorities within the administration. Although the observations in this particular section are drawn from the actual events in that college, these remarks may apply to most programs elsewhere, enabling us to draw general conclusions.

The first factor that precipitated the decline of the program was a perceptible shift in the needs of the college's service area. As desktop publishing began to decline and was replaced by digital communication, jobs in these new areas became "hot" and relevant. As the college is in a service area that feeds the city's largest newspaper group, jobs in digital communication began to increase exponentially. At the same time, faculty to teach these programs became available since, as a workforce program, faculty needed to have work experience more than academic credentials. It was possible to find qualified, experienced people to teach digital art, for example, because work experience took precedence over degrees. Finding people to teach technical communication was not as easy; practitioners were making good money in the 1990s and did not want to teach for low salaries. Although the program did have one highly qualified person in technical communication, the pressures of teaching a full course load of five classes a semester prevented any attention to program growth. As a result, the program languished. Changing priorities within the administration created a push toward programs that would offer job placements immediately and efficiently to a large number of students; this directly affected program ratings because workforce programs can be rated exemplary only if a certain percentage of students obtain employment after completing the program. As the needs of the service area gravitated toward digital communication, priorities shifted toward that area because jobs and internships were easily accessible. The absence of a dedicated program administrator for technical communication hastened the decline of technical communication programs even further.

In contrast, in the mid-2000s, in a different part of the city, the college saw an increasing demand for the technical writing course. Inquiry showed that the demand was coming from students in the Drafting and AutoCAD programs; technical writing was a required class for these majors. Students from this program were (and continue to be) in great demand. Other workforce programs were established, such as Piping and Drafting. These programs aimed to provide qualified and trained workers to staff the booming oil and gas industry. In addition, technical personnel were being recruited in large numbers for offshore projects in Iraq and other countries. All these programs required a number of math courses such as College Algebra and Trigonometry; however, Technical Writing was the only required English course. With a total of 16 programs in this area, the demand for the class has not only been increasing but has also served to raise the profile of the technical writing course from a service course to an integral program requirement.

This factor has served as the main driving force behind attempts to revive and re-create a program in technical communication; this time, the service-area needs and current economic conditions are playing a major role in determining the structure and form of the program. In subsequent sections, I discuss the various factors and drivers that have played an instrumental role in shaping perspectives and given form to the program.

## FOCUS GROUP FINDINGS

As an academic, I was well aware of the gap between theory and praxis. Brockman, as early as 1979, had discussed the need for a close working relationship with industry as a key requirement for the success of a technical communication program.

To understand employer needs and tailor our offerings to industry expectations, I conducted a focus group attended primarily by office bearers of our local STC chapter. Some attendees were, and still are, hiring managers in large corporations. The feedback that emerged from the focus group was insightful and, in a couple of instances, contrary to existing scholarship. Several scholars write about the dichotomy between industry and academics, going so far as to discuss the "clash of academic and corporate cultures" (Kratz, 2000); however, our focus group was almost unanimous in the opinion that this divide needs to be bridged in the interest of the profession. The focus group offered the following suggestions:

- Most participants agreed that some of their technical writers came in from other disciplines or from other job functions and learned technical writing on the job. They all felt that formal technical writing credentials would greatly improve the caliber of technical writers.
- Further, most participants opined that a certificate or associates degree from a 2-year institution would help equip entry-level technical writers. They agreed that they would employ certificate-holders for these positions. In fact, two of the hiring managers felt that a certificate or associates degree in technical writing was preferable to a bachelor's degree in another discipline.
- Interestingly, most of the hiring managers felt that it would be useful for students to possess basic knowledge of the business world. To this end, one participant suggested that a basic, survey-type course in business practices would be extremely beneficial.
- Most participants felt that too much emphasis was placed on expertise in tools and software such as RoboHelp, FrameMaker, and InDesign. The general opinion was that it was more important for entry-level technical writers to understand the rhetorical context and to be able to engage in an effective audience and task analysis. As one participant said, "All I want is for them to be able to write effectively; I can teach them the tools, that's

easy." This opinion contrasts with a previous report, dated 2000, suggesting that "employers say they need someone with experience in RoboHelp, FrameMaker, and so on and are initially satisfied that they fulfilled the requirement for a manual, online help, etc." (Tews, Johnson, & van Oss, 2000). Participants agreed that knowledge of writing contexts, understanding of the audience and purpose, and appreciation of the rhetorical rationale in document design were significant skills. They also felt that they were not in a position to teach these concepts and theories.

- Finally, most focus group contributors felt that the key ingredient missing in their existing technical writers, and the skill that was most difficult for them to impart on the job, was audience and task analysis. All participants agreed unanimously that audience and task analysis was something that needed to be taught in a classroom using a theoretical framework. One hiring manager felt that, as long as technical writers were able to analyze the purpose and users of a document, they would be able to learn the details of producing such documents quite easily. It was interesting that, although the terms *rhetoric* or *rhetorical analysis* were not used, all participants described variations of them. This input, combined with the opinion that knowledge of tools was only secondary, gives us a clearer picture of the importance of fundamental theoretical concepts; further, we realize that these concepts can be put in place only by a strong, foundational course in technical communication.

The focus group's inputs gave us a clear picture of the actual skills and concepts that hiring managers were seeking in entry-level and experienced technical writers; their feedback offered not only clues into the kinds of courses we needed to offer but also into the nature of these courses. We understood that we needed to offer theoretically sound courses that combined mastery of concepts with ample opportunities for the application of these concepts.

## Identifying Service Area Needs

The first step in setting up a program in technical communication was to identify the needs of our service area. We realized that offering a customized program tailored to the needs of area employers would increase the value and credibility of the program. Our vision was to create a program that would "prepare students to meet the productive communication demands of our increasingly sophisticated technological society" (Coppola, 1999, p. 251).

Our city focuses on three major industries: oil and gas, energy, and chemicals. It also offers expansive services in the medical and financial sectors. However, the location of our campus in the energy corridor of the city meant that we needed to offer courses tailored to the energy, oil, and gas sectors. Discussions with these employers resulted in feedback similar to the input from the focus group. Industry sources requested the same skillsets: basic writing skills; the ability to produce coherent, usable documents; and an understanding of business writing. Another requirement that came up frequently was the ability to understand and adhere to a style guide. Oil and gas companies usually create style guides used extensively in requests for proposals (RFPs), statements of work (SOWs), and service-level agreements. A number of respondents felt that a basic understanding of style guides and some exposure to their use would be most beneficial. A small number of respondents also felt that it would be useful to familiarize students with rudimentary terminology relevant to the oil and gas industry. For example, one hiring manager at a major oil company mentioned that it would be useful if potential technical writers could obtain an overview of significant terms such as *upstream*, *downstream*, and *midstream*.

Similar needs seemed to exist in the medical field as well; the requirement seemed to be for the ability to write various kinds of documents with a need for some specialized knowledge. Because the college already offers courses in medical technology and paramedical skills, a class in basic medical terminology has already been developed and can be included in the technical communication program as an elective. Besides medical terminology, the college also offers classes in medical office procedures and other related areas that could be offered to potential technical writers in the medical field.

## TRANSFER OF CREDITS/INTEGRATING WITH BACHELOR'S PROGRAMS

One of the many attractive features of community college programs is the ability to transfer credits to receptive 4-year institutions. With the increasing cost of tuition in 4-year institutions, increasing numbers of students enroll in community colleges to complete first-year and sophomore courses. Recognizing this, our college has structured articulation agreements with universities in the area to facilitate seamless transfer of credits. Studies have shown that "variables affecting transfer included the distance from the college to the nearest 4-year institution, the academic preparedness of incoming high school students, and household income of the county in which the college is located" (van Ommeren, Liddicoat, & Hom, 2008). However, in the case of the technical communication program under consideration, other considerations affected this seamless transfer; I will discuss these factors in this section.

Technical communication programs at most 4-year institutions follow a model in which first-year and sophomore classes are mainly drawn from the core curriculum of the state. Junior- and senior-level classes offer specialized courses toward the major (and minor, if any). This can be clearly seen from the following screenshot of the course listings at a 4-year university in the area (see Figure 1).

Because community colleges are allowed to offer only first- and second-year classes for academic credit, any technical communication program proposed in a

**COMMON CORE REQUIREMENTS (42 HOURS)**

See listing under General Education Requirements in the College of Humanities and Social Sciences. COMM 3306 is recommended in fulfillment of the speech communication component.

**PREPARATORY REQUIREMENTS (9 HOURS)**

Philosophy (3 hours Ethics) Upper Level elective (3 hours) Lower-level electives (3 hours)

**UPPER-LEVEL REQUIREMENTS (45 HOURS)**

ENG 3302 Business and Technical Report Writing
ENG 3316 (History and Rhetoric) or ENG 3317 (Studies in the Theory of Rhetoric)
ENG 3330 Desktop Publishing
ENG 3331 Advanced Desktop Publishing or ENG 3335 Publishing for the Web
ENG 4099 Professional Writing Portfolio. Required in semester of graduation (0 hours)
ENG 4322 Editing, Rewriting and Copyreading (prerequisite ENG 3318) (3 hours)
ENG 4380 Field Experience

**OTHER REQUIREMENTS (27 HOURS)**

Select 7-9 writing or publishing courses (21-27 hours) from:
ENG 3304 Advanced Business and Technical Report Writing
ENG 3308 Legal Writing
ENG 3325 Medical Writing
ENG 3326 Proposal Writing
ENG 3328 Documentation and Manuals
ENG 3329 Environmental Writing
ENG 3331 Advanced Desktop Publishing or ENG 3335 Web Publishing
ENG 3333 Writing for the Media (or COMM 3311)
ENG 3334 Writing for Presentation (or COMM 3312)
ENG 3336 Writing for the Web
ENG 3342 Organizational Publications
ENG 3371 Public Relations (or COMM 3301)
ENG 3373 Introduction to Advertising (or COMM 3303)

Figure 1. Course listing from a nearby 4-year institution.
**Source**: http://www.uhd.edu/academic/colleges/humanities/english/pw.html

community college setting needs to recognize the problems of transferring credit hours. Existing programs in technical communication in 2-year institutions offer specialized courses as workforce courses as laid down in the Workforce Education Course Manual (2010–2011). A close examination of the following screenshot shows that although some courses are listed as "SCH" (semester credit hour), others are designated as continuing education credits (CEU) (see Figure 2).

Although academic credit hours transfer easily, we found that most of the courses listed as workforce courses (with course prefix ETWR in the previous table) were not as easy to transfer. The University of Texas website, for example, clearly stated that

> in addition to coursework parallel to the first two years of university study, community colleges offer curricula designed to provide students with immediately marketable career skills. Such workforce (also known as *occupational*, *technical*, or *vocational*) courses are not designed for transfer into university curricula and, with only a few exceptions, are *nontransferable* at UT Austin. (http://www.utexas.edu/student/admissions/ate/problems/workforce.htm)

Most courses required for a technical communication program are necessarily workforce courses, because they do not have equivalents in the *Lower-Division Academic Course Guide Manual* (ACGM, the academic counterpart to WECM) as transferable credit courses. In our program, for instance, the only ETWR course cross-listed as an academic course is ETWR 2301, also listed as ENGL 2311 in the ACGM. As the previous table shows, many other courses, including courses with nonspecific descriptions, are mostly workforce courses and therefore difficult to transfer. On the one hand, it is advantageous to have courses with general titles such as Intermediate or Advanced Technical Writing because these can be adapted to local service-area needs; for example, an Intermediate class could be designed as a "special topics" class that would be tailored to specific industry needs. On the other hand, however, students who wish to transfer to a 4-year university for an undergraduate degree may question the utility of these courses if they do not transfer easily.

One possible solution to this problem would be to create special, clearly worded articulation agreements with selected 4-year institutions. Our early and preliminary dialogs, especially with programs that offer bachelor's and master's programs, showed their willingness to enter into articulation agreements. No doubt, close evaluation of our syllabi and course outcomes for these courses would be a precursor to such articulation. What interested us, however, was the reasoning behind this rather unexpected acceptance; as one program director put it, the students who entered their undergraduate program were not really prepared for technical communication; they just completed the basic, core curriculum requirements anyway. In most cases, their only exposure to the field was the

| CIP | Rubric Number | Course Title | Status | Range | Course Type |
|---|---|---|---|---|---|
| 23.1303 | ETWR 2301 | Advanced Technical Writing | Active | 48-96 | SCH |
| 23.1303 | ETWR 1043 | Technical Writing II | Active | 7-40 | CEU |
| 23.1303 | ETWR 1191<br>ETWR 1291<br>ETWR 1391<br>ETWR 1491 | Special Topics in Professional, Technical, Business, and Scientific Writing | Active | 16-112 | SCH |
| 23.1303 | ETWR 1x64<br>ETWR 1x65<br>ETWR 2x64<br>ETWR 2x65 | Practicum (or Field Experience) - Professional, Technical, Business, and Scientific Writing | Active | 112-640 | SCH |
| 23.1303 | ETWR 1091 | Special Topics in Professional, Technical, Business, and Scientific Writing | Active | 7-112 | CEU |
| 23.1303 | ETWR 1000 | Grant Writing | Active | 7-40 | CEU |
| 23.1303 | ETWR 2303 | On-line Documentation | Active | 48-96 | SCH |
| 23.1303 | ETWR 2003 | On-line Documentation | Active | 48-96 | CEU |
| 23.1303 | ETWR 1302 | Introduction to Technical Writing | Active | 48-96 | SCH |
| 23.1303 | ETWR 1002 | Introduction to Technical Writing | Active | 48-96 | CEU |
| 23.1303 | ETWR 2305 | Intermediate Technical Writing | Active | 48-96 | SCH |
| 23.1303 | ETWR 2405 | Intermediate Technical Writing | Active | 64-96 | SCH |
| 23.1303 | ETWR 2005 | Intermediate Technical Writing | Active | 48-96 | CEU |
| 23.1303 | ETWR 1005 | Technical Writing I | Active | 7-40 | CEU |
| 23.1303 | ETWR 2001 | Advanced Technical Writing | Active | 48-96 | CEU |

Figure 2. Course designations.
**Source**: http://www.thecb.state.tx.us/aar/undergraduateed/workforceed/wecm/wecmlist.cfm

sophomore technical writing course. The program director felt that students from a workforce program, albeit one with nontransferable credits, would have a better foundation in technical communication. We realized that, for our program to succeed and for students to enter the program confident about its transferability, we would have to create clear, detailed articulation agreements with 4-year institutions to facilitate seamless transfer of credits.

Another possible solution lies in the route taken by some community colleges and universities to create an articulation agreement from transfer from a vocational or workforce program to a baccalaureate program in an applied area. For example, some universities offer programs such as the Bachelor of Applied Science (BAAS), Bachelor of Applied Technology (BAT), and the Bachelor of Science in Nursing (BSN); it is possible to transfer workforce credits to these programs. The Texas Two-Step Project (Technology Workforce Opportunities through Seamless Transitions and Educational Partnerships) allows for students taking technical and vocational courses to transfer seamlessly from high school to community college in the first step and then from community college to a 4-year university in the second. At the time of writing, no technical communication programs seem to be offered under this program; however, this would be a feasible option for workforce-driven programs in technical communication.

In sum, a number of course options are available for a technical communication program in community college; although there are real problems and concerns with transfer of credit, these are not insurmountable and can be solved. Students and industry will benefit from such resolution.

## ROLE OF THE TECHNICAL WRITING COURSE IN COMMUNITY COLLEGES

In general, technical writing (designated as ENGL 2311 in most Texas public institutions) is regarded as a "service course." Although it meets the sophomore English requirement for the Core Curriculum, few students in our college system take it as such. Most students, in my experience, take a literature class from among the many offered, ranging from American and British literature to Mexican-American and multicultural literature, to meet their sophomore English requirement. The general impression is that technical writing is not really an elective of choice for the sophomore requirement.

This impression is compounded by the fact that many of our workforce programs, from Audio Engineering to Nursing, specify technical writing as a required course. This requirement gives not only students but also academic counselors and advisors the impression that technical writing is some sort of "vocational" course (especially because so many workforce programs require it). This impression creates a conundrum: technical writing classes fill up quickly while many literature classes see poor enrollments, sometimes cancelled because they do not "make." Technical writing classes always fill to capacity days after registration begins; yet these classes are full of students who take it because they *have to*, not because they want to. Thus, the tenor of a technical writing class is influenced by both of these impressions—an administrative perception of it as a vocational rather than an academic course and the student perception of it being an imposition rather than a choice.

The prerequisites for technical writing as opposed to literature further complicate the conduct of the class. To take a literature class, students must have completed two courses (6 hours) in composition; the second of these involves writing an academic research paper. Thus, when students enter a literature class, they are well prepared for research, synthesis, paraphrasing, and complex writing. However, students wishing or needing to take technical writing need only one composition class as a prerequisite. Thus, they have not, in most cases, obtained any exposure to the rigors of the research paper; nor have they done as much writing as their counterparts who take literature classes. Moreover, as workforce students, they perceive writing as difficult, mysterious, and stressful; they often voice frustration at being good in their technical field but having to struggle with writing. Thus, they enter the class with resentment that quickly turns to apprehension when they realize how much writing they will need to do to pass the course.

The solution to this problem is twofold: first, we need to ensure that students required to take technical writing are better prepared; this could be accomplished either by requiring them to take a second composition class or by designing a special Composition I class that would provide them with better writing skills. Second, we need to raise the profile of technical writing to that of a sophomore elective on par with literature; this also requires us to emphasize that technical writing is not just for workforce students but proves extremely useful in all business writing as well. Nagelhout (1999) stated that the "technical writing class, then, should be a vehicle for students to reflect on the ways that writing functions in their discipline, as well as the ways that writing functions in our society" (p. 286). Thus, a repositioning of the technical writing class as a class that teaches business writing and therefore fulfills a broader purpose would go a long way in changing the profile of the course.

## CONCLUSION

Our needs-assessment study and research showed that many of Brockman's observations in 1979 still prevail in certain areas. Brockman's emphasis on close relationships with industry holds true; many of our focus group participants appreciated the fact that we were seeking their input. They not only felt heard, but also became invested in helping our program succeed. This was evident from their willingness to provide internships, to serve on an advisory board in the event of one being formed, and to mentor students. The 2-year duration of community college programs did not come across as a major disadvantage as long as credits could be transferred. The open admissions policy brings with it varying levels of preparedness among students; this factor plays a vital role in student success and, as a corollary, on program success. Since 1979, technical communication programs have gained credibility and acceptance; therefore, Brockman's indication of "newness" may no longer be as much concern.

## REFERENCES

Brockman, J. B. (1979). *Troubleshooting in technical communication programs in community colleges* (pp. 2-18). ERIC document No. (ED179975).

Coppola, N. (1999). Setting the discourse community: Tasks and assessment for the new technical communication service course. *Technical Communication Quarterly, 8*(3), 249–267.

Kratz, S. (2000). The dual mission of the community college and implications for technical writing instruction. *CPTSC proceedings.* Retrieved from http://cptsc.org/pro.html

Nagelhout, E. (1999). Pre-professional practices in the technical writing classroom: Promoting multiple literacies through research. *Technical Communication Quarterly, 8*(3), 285–299.

Tews, D., Johnson, R., & van Oss, J. (2000).What industry looks for in technical communication graduates. *CPTSC Proceedings.*

van Ommeren, A., Liddicoat, C., & Hom, W. (2008). Developing service area indices for community colleges: California's method and experience. *Community College Journal of Research and Practice, 32,* 463–479.

CHAPTER 8

# Globalizing Technical Communication Programs: A Diachronic Perspective

***Laurence José***

Initiating a conversation on globalization and its impact on scientific and technical communication programs can be intimidating, if not overwhelming. Indeed, besides its popularity, its rampant usage in the academy, and the fast-growing research it has enacted, globalization is also in many ways a semantico-referential "riddle" that involves a plethora of different aspects (technological, historical, cultural, linguistic, political, social, ethical, just to name a few).[1] As a corollary, globalization offers many research possibilities that can easily span across disciplinary borders. Without negating the multifaceted dimension of the topic, the present chapter focuses on a very specific characteristic, that is, the programmatic weight of the global context and its significance for defining the mission of technical communication programs. More specifically, this chapter grapples with such questions as

[1] The origin of the term itself is contested. Some, like Joel Spring (2009), contend that "the term 'globalization' was coined by the economist Theodore Levitt in 1985 to describe changes in global economics affecting production, consumption, and investment" (p. 2); and others state that the term was used well before 1985. As the *New York Times* noted in Levitt's obituary, "While Mr.Levitt's work was closely associated with the idea of globalization in economics, and while he published a respected paper in 1983 popularizing the term, he did not coin the word. (It was in use at least as early as 1944 in other senses and was used by others in discussing economics at least as early as 1981.)" (*New York Times,* July 6, 2006).

- What exactly is the programmatic weight of globalization in technical communication? How is it manifested and what is the history of programmatic work in this domain?
- What are the larger implications and the significance of positioning programs in the global context?
- How do the current challenges (induced by the global context) speak to the history of the field and the mission of technical communication programs?
- What strategies should guide the positioning of technical communication programs in the global context?

By addressing these questions, I seek to provide a comprehensive overview of the development of the global context as a programmatic theme in technical communication while exploring the meaning of its history in light of the current contingencies. My goal is to make way for productive conversations about the exigencies of the current context for technical communication program administrators. Specifically, this chapter is rooted in the contention that a diachronic approach is a means to gain a better understanding of the implications of the current challenges and of their affordances for defining the mission of technical communication programs. In this sense, this chapter not only extends an ongoing conversation in the field—that is, how to define strategies for effectively positioning programs in the global context—but also highlights how the current exigencies offer a locus for reflecting on and for reconciling the demands of the global context with the field's larger missions.

Before "taking a look back" and embarking on a diachronic journey, I would like to begin this chapter by offering some more context to frame the discussion and most importantly, to justify my stance regarding the goal and the timeliness of a diachronic examination.

## FIRST, SOME CONTEXT

There is nothing new or original in stating that the global economy directly impacts the profession of technical communicators. More and more, technical communicators have to adapt their communication practices to the needs of a linguistically and culturally diverse audience; new technologies as well as international business partnerships facilitate the rise of communication processes above national and cultural borders and consequently call for communication practices that can effectively cross these borders. These exigencies have greatly influenced the research in the field during the last 30 years and have fueled a fast-growing scholarship such as, Deborah Andrews (1996), Linda Beamer (1992), Nancy Hoft (1995), Thomas Huckin (2003), Bruce Maylath and Jeffrey Grabill (2009), Natalia Matveeva (2006), Emily Thrush (1997), Doreen Starke-Meyerring and Melanie Wilson (2008), and Timothy Weiss (1995). In

1999, at the turn of the new century, Carl Lovitt and Dixie Goswami noted that "the explosion of interest in international professional communication represents one of the most dramatic developments in the fields of business and technical communication over the past decade" (p. 1). Since then, the interest has not only been sustained, but it has grown exponentially. The amount of literature dedicated to the topic during the last 20 years conjugated with the emergence of global partnership projects and other pedagogical initiatives (Starke-Meyerring & Wilson, 2008) offer a perfect illustration of how much globalization has impacted the research in technical communication.

Given the intimate relationship between technical communication and the workplace, issues of globalization have naturally become a major programmatic concern and have entered the realm of program administrators' discourses. In 2002, more than 10 years ago, Thomas Huckin stated in his keynote address at the conference of the Council for Programs of Technical and Scientific Communication (CPTSC), "Most graduates of our [technical and professional communication] course go on to careers in American corporations, and most will either work abroad for some time or have significant interactions with colleagues or customers overseas" (p. 6). In the last 8 years, the programmatic weight of globalization has been more widely recognized and emphasized. In 2008, Starke-Meyerring and Wilson referred to this deep and long-lasting impact of globalization on program and curriculum design to introduce their work on *Designing Globally Networked Learning Environments*:

> There is perhaps not a single day that faculty, administrators, international program experts, and many others in higher education are not inundated with news about globalization with such title phrases as "world changing," "discontent," "survival," "global marketization of higher education," "global trade," "or higher education under siege," signaling the highly contested and charged nature of the discourse surrounding these terms. Often, this contested discourse is accompanied by growing pressure and urgent calls on faculty and administrators to act quickly—to "internationalize" their campuses, curricula, and classrooms; to position their institutions, programs, and courses in globalizing higher education environments and "global markets." (p. 1)

In addition to underscoring the significance of the global context for administrators, Starke-Meyerring and Wilson's (2008) statement points to another key feature of globalization: the quasi-rhetorical meaning the term has acquired in the academic context. Indeed, invoking globalization at a programmatic and curricular level rhymes with money and serves specific agenda and goals. In a context in which the demands to internationalize and respond to the global context are so powerful and engage budget related considerations, it is often difficult to remain critical vis-à-vis globalization as a phenomenon, despite (or maybe because of) all the social, political, ethical, and cultural issues it raises. Its ubiquitous presence, and some may say its "catchphrase" dimension, combined with the recent and fast-growing character of its scholarship, are all factors that greatly complicate the assessment of what globalization has come to mean and to imply for program administrators.

The recognition of the challenges underlying the positioning of technical communication programs in the global context are not new and have been largely emphasized in the literature. However, today's educational and economic landscape calls for engaging in a reflection on the history and development of globalization as a programmatic theme. Indeed, we are today in a context in which the pressures to position programs in the global context are concurrent with a worldwide economic crisis that tends to prompt isolationist reflexes and with the emergences of voices that directly question the function and usefulness of international processes in higher education. For instance, in June 2011, *Inside HigherEd* published a piece by Elizabeth Redden titled, "The End of Internationalization?" This (somewhat) provocative question is in fact eponymous of the title of one of the sessions from the 2011 NAFSA: Association of International Educators conference. In the article, Redden described some of the queries and also the doubts voiced by educators regarding the worth and the end of internationalization in higher education:

> Panelists at the session argued that as internationalization has moved from a fringe to core university activity, it remains imperative for professionals to scrutinize what they're working toward and why. Every university now says it wants to be "international," but what does that mean—as [Jane Knight, a professor at the Ontario Institute for Studies in Education at the University of Toronto] said, "internationalization has become a catch-all phrase for everything"—and to what end?

The comments made by presenters at the NAFSA conference, which brings together international education professionals, illustrate the tensions educators and administrators are facing: on one hand, the pressures to "internationalize" —and "globalize" are not subsiding—it is a "a core university activity" (Redden, 2011)—and on the other hand, skepticism and questions regarding the "why" are on the rise.[2] Maybe, as Redden contended, this is an indication that internationalization is suffering from a "mid-life crisis." But regardless if one agrees or not with the "mid-life crisis" qualifier, the conversations held by the very promoters of international education clearly indicate the need and the necessity for educators and administrators to take a step back and consider the end of

[2] On May 5, 2011, *The Chronicle of Higher Education* published a piece by Mark A. Montgomery that addresses very strong critiques toward the current international education system in the United States. This article elicited a response by Ann Imlah Schneider on June 1, 2011.

internationalization and other global initiatives in conjunction with the larger goals and mission of higher education programs. Given the concerns of technical communication program administrators for responding to the challenges of the global context, I contend that the current debates offer a perfect argument for adopting a more time-encompassing stance on what globalization has come to mean and imply. In other words, the synchronic conversations revolving around internationalization in higher education constitute an opportune moment to engage in a diachronic and reflective exploration of global issues as a programmatic theme in technical communication.

To begin this query, I propose to examine how globalization has progressively infiltrated the discourse of technical communication program administrators.

## GLOBAL ISSUES AS DISCURSIVE ENTITIES: A RETROSPECTIVE

The role globalization has come to play at a programmatic level is directly visible in program administrators' discourse. In this regard, the list of the grant winners designated by the CPTSC as well the presentations given during its yearly conference are particularly revealing, as they offer a direct reflection of what administrators value and need over a specific time span.

### Research Grants

The CPTSC research grant program aims at encouraging and promoting programmatic research. It is a relatively recent program—the first grant was awarded in 2003 (Maylath & Grabill, 2009, p. 34). However, despite its temporal and hence epistemological limitation, this list is still a good instrument to characterize the needs and goals of programs over a period of time. With regard to the significance of globalization, a brief examination of the grant winners' titles between 2003 and 2010 reveals at least two things:

- Issues surrounding globalization are part of program administrators' discussion (between 2004 and 2010, out of 17 project grants distributed, two sponsored projects are explicitly related to globalization).[3]
- The conversations surrounding the programmatic meaning of globalization are distributed over a period of time and are not closed; the first project on globalization was sponsored in 2004 and the second project was sponsored 3 years later, in 2007.

[3] Although it does not directly invoke globalization as a theme, Michael Salvo's 2009 sponsored research project on "European Developments in User-Centered Design: Articulating Centers of Practice" exemplifies the internationalization of programmatic research (Maylath & Grabill, 2009).

If the presence of global issues in programmatic discussions is not surprising, it is still worth noticing that the topic is of sustained interest and that it can be approached from different angles and raises issues with different ramifications and meanings for technical communication programs. To illustrate these two points, let's take a closer look at the two projects on globalization that the CPTSC chose to sponsor:

- The first project, "Global Programs Partnerships in Technical Communication," presented in 2004 by Doreen Starke-Meyerring and Ann Hill Duin, is based on a Web-based survey and aimed at exploring emerging trends in global partnership development.
- The second project, "Linguistic and Cultural Diversity in Scientific and Technical Communication: Designing International Curricula," presented in 2007 by Ann Brady and Laurence José, involves curricular research and aims at defining new means to internationalize curricula.

Despite their similar global and programmatic orientations, these two projects offer an interesting demonstration of the variable issues globalization raises and its different levels of application.

Although the CPTSC grant winners list provides insightful data for assessing the currency and multifaceted status of globalization in programmatic research, the youngness of the research grant program does not really allow one to fully gauge the development and growing weight of global issues for program administrators throughout the years. To overcome, or to at least attenuate, these epistemological limitations, I propose to intersect the findings from the CPTSC grant winners list analysis with an examination of the CPTSC annual conference program. To that end, I analyzed the conference's programs from 1974 to 2010.

## CONFERENCES

To gather the data, I downloaded all the proceedings from the CPTSC proceedings database (http://www.cptsc.org/pro.html).[4] Given that the CPTSC proceedings database "only" spans from 1974 to 2008, I performed a web search using Google to find the online versions of the conference program for the years 2009 (http://www.asb.dk/article.aspx?pid=21482) and 2010 (http://cptsc2010.wordpress.com/program/). Once I had collected the data, I queried the

[4] The proceedings from 1974 to 2009 are available from the CPTSC database with the following exceptions: no proceedings were published in 1976. The 2011 and 2012 CPTSC conference programs were not yet available when I queried the database. Although they weren't included in the results, a quick overview of these programs confirms the growing weight of global themes as a programmatic issue.

following terms and phrases in the titles of the conference presentations: global, globalization, intercultural/multicultural communication, and international communication.[5] After perusing the table of contents of the proceedings and reading the online programs for the years 2007, 2008, and 2009, I established that

- 1993 was the first year international issues were explicitly referred to in a presentation title. Indeed, that year Deborah Andrews gave a presentation on "International Dimension of Technical Communication" (the presentation was part of a panel on "Curriculum Design: The Workplace and the Academy").
- 1996 was the first year the adjective "global" appeared in a title and the first year a whole panel was dedicated to global issues. In other words, 1996 was the first year that globalization served as a theme to categorize and to delineate a cluster of programmatic research. Specifically, that year the conference program included a panel on "Remapping Programs, Local and Global Communities: Influences and Options."
- Except for 1995, there has always been at least one presentation pertaining to global/international issues since 1993.
- In 1996, 1997, 1998, 1999, 2001, 2005, 2006, 2007, 2008, and 2009, the conference program included entire sections and panels focusing on global-related issues.
- In 2004, a presentation addressing global issues was part of a plenary panel.
- Both in 2002 and 2006, globalization was the central theme of the keynote address.

If the data aggregated through the excavation of the CPTSC annual conference programs and proceedings corroborate the findings regarding the central status of global issues from a programmatic perspective, it also underscores the young and fast-emerging character of these issues as well as their sustained and growing presence. It is remarkable to note that, within just 17 years, the conference went from one isolated presentation on international issues in 1993 to presentations constituting panels or included in plenary panels and to keynote presentations explicitly focusing on and organized around global-related questions. Since 1996, when the word "global" appeared in a presentation title for the first time, globalization has become the focal point of an ongoing conversation in the agenda of program administrators. Perusing the CPTSC conference program over the last 38 years illustrates not only the field's endeavor to shape programs

[5] The use of the adjectives "international, intercultural/multicultural" as fitting epithets for my query is based on the fact that globalization enacts processes that cross linguistic and cultural borders. Although I employed these query terms as a way to initiate the analysis, I also completed my investigation by reading individually each title in order to more precisely assess its eventual global dimension.

according to the exigencies of the workplace, but it also epitomizes the weight globalization has acquired in terms of programmatic considerations and decisions. In addition to meeting the demands of the 21st-century workplace, it is important to note that the CPTSC's commitment to expand and to encourage reflection around the programmatic and curricular dimension of globalization can also be seen as part of a larger commitment to include diversity within the scope of technical and professional communication.[6]

To pursue the assessment of the global context's programmatic weight, I propose now to shift the analysis from a discursive level to a more referential level by exploring how global issues have been translated in the shaping of technical communication programs and curricula.

## CONSIDERING THE REFERENTIAL MANIFESTATIONS

When transposed at a referential level, an investigation of global issues as a programmatic theme seems to confirm its increasing weight and impact. Indeed, in recent years, several universities have created program certificates with an explicit focus in international technical communication. For instance,

- Bowling Green State University offers an online graduate certificate in international scientific and technical communication;
- Kennesaw State University offers a graduate certificate in professional writing for international audiences;
- The University of Wisconsin-Milwaukee offers a graduate certificate in international technical communication.

Considering the general vocational aspect of program certificates, the existence of certificates in international technical communication shows that being able to communicate with international audiences is reckoned as a skill constitutive of today's professional communicator's identity. Moreover, the creation of such program certificates underscores the extent to which globalization informs and impacts programmatic and curricular discourses. In other words, it is a direct manifestation of how much the exigencies underlying globalization have become a strategic part of programmatic planning. As I am reflecting on the significance of such certificates for characterizing the programmatic impact of the global context, it is important to note that the emergence of certificates in international

[6] The CPTSC Graduate Student Diversity Scholarship, created in 2008 and sponsored by Bedford/St. Martin's, is yet another manifestation of this commitment to diversify the context and discourse of professional and technical communication in the academy. The goal of this scholarship is to "increase diversity in its membership and in the field of technical communication. Graduate students of color, race, ethnicity, disability, or nationality who are underrepresented in the technical communication field and who are enrolled in a technical communication graduate program are eligible to apply for this scholarship." (CPTSC, 2008)

technical communication can also be seen as a way to better characterize the overall rationale of program certificates in technical communication. For instance, in his doctoral work on program certificates, Jim Nugent (2009) considered that the very existence of certificates with an emphasis on international technical communication is a means to highlight the flexibility and concern for local needs characterizing program certificates (p. 116). As he suggested, certificates such as the ones offered by Bowling Green State University, Kennesaw State University, and University of Milwaukee are means for complicating our understanding of program certificates and of their role in meeting the needs of the workplace.

Albeit the influence of globalization in the creation of program certificates dedicated to communicating in linguistically and culturally diverse contexts, the broader impact of globalization on programs and curricula seems to be, if not less visible, at least more challenging to determine. Hence, in their 2005 study, "Trends in Scientific and Technical Communication Programs," Sandi Harner and Anne Rich found that only 1% of undergraduate programs in technical communication require a course explicitly on the topic of global or international communication, and only 5% of programs offered such a course as an elective. Of the 80 programs they examined, 7 included a modern language component as a part of the curriculum, a requirement that typically stands as the only distinction between the bachelor of science and the bachelor of arts degrees (pp. 215–216).[7] As I write this, 6 years removed from when the article was published, Harner and Rich's work remains arguably one of the most in-depth and reliable sources for data on the curricula of technical communication undergraduate programs.[8] Studies on academic programs in technical communication are very scarce and, in this very limited ensemble, Harner and Rich's article constitutes one of the most recent and complete studies on undergraduate programs in technical communication.

Despite its catalytic effect for the creation of specialized program certificates, it appears though—and this may be a surprise considering the popularity and ubiquity of the topic—that technical communication programs and curricula offer little reflection of the programmatic weight globalization has acquired throughout the last 30 years. However, and to slightly nuance this last statement, the difficulty in assessing the programmatic impact of globalization can also be attributed to the fact that delineating samples for research on academic programs in

[7] A foreign language requirement can be considered as a programmatic means for emphasizing the importance of being able to communicate across linguistic and cultural borders for today's technical communicators.

[8] For their study, Harner and Rich (2005) determined their program sample by using the STC Academic Programs Database. The recent work of Jim Nugent (2013) and Lisa Meloncon (2012) offer new and productive data to understand technical communication programs. Both studies were published after the research for this chapter was completed.

technical communication is challenging and is not always a transparent or clearly explained process in studies on academic programs. As Nugent (2009) noted,

> Most of [the] studies on academic programs draw at least some portion of their sample from a major program directory, and in doing so, make assumptions (whether articulated or not) about the completeness and representativeness of those directories. Each study also makes assumptions about "what counts" as a program for the purposes of their research, and assumes that the programs that "count" are represented in their samples. These assumptions, and the diverse methods these studies ultimately employ, demonstrate just how challenging sample determination is for research on academic programs in technical communication. (p. 79)

Notwithstanding the challenges one might encounter to provide an exact account of the programmatic weight of the global context at a referential level, a retrospective analysis of the discursive and referential manifestations of globalization in the history and shaping of technical communication programs shows that

- The global context has undeniably acquired a programmatic dimension; this has been particularly notable since the early 1990s.
- Defining strategies for positioning programs and curricula in the global context remains a major component of today's conversations and counts as an important part of program administrators' agenda. The sustained interest of program administrators for research projects on globalization underscores the persistence of a gap at this level.
- A discrepancy exists between the discursive manifestations of global issues and their translation into visible/tangible programmatic changes.

If such observations undeniably allow for a more nuanced approach and carry important lessons for programmatic planning, I believe that it is essential to intersect these factors with the implications and ramifications underlying the enactment of responses to globalization. To fully assess what globalization has come to mean at a programmatic level, I propose to now shift the perspective from "looking" at occurrences of globalization to "listening" to the voices of scholars who best exemplify the larger ramifications underlying the positioning of technical communication programs in the global context.

## IDENTIFYING THE AFFORDANCES OF THE GLOBAL CONTEXT

During the last 10 years, the impact of globalization on technical communication programs has become the locus for discussions about its sociopolitical ramifications. Indeed, following the exponential growth of global issues on the programmatic scene in the mid-1990s, some scholars have, at the turn of the millennium, called for the necessity to pay greater attention to the significance of the sociopolitical and ethical aspects of globalization. In this regard, Thomas Huckin's 2002 CPTSC keynote address certainly counts as a milestone. In front of an audience composed of technical communication scholars and administrators, and following his emphasis on the prominent status of globalization, he forcefully argued for the development of a critical approach:

> At a time when both US imperialism and corporate America are becoming more and more dominant and are under intense criticism both at home and (especially) abroad, there is a glaring absence of global sociopolitical consciousness in TPC [Technical and Professional Communication]. I think this is a major problem. (pp. 7–8)

In the following years, key scholars in international and intercultural technical communication by referring explicitly to his argument, extended Huckin's (2002) call and initiated conversations for deepening the reflection on the social, political, and ethical consequences of globalization. For instance, in 2005, in "Meeting the Challenges of Globalization: A Framework for Global Literacies in Professional Communication Programs," Doreen Starke-Meyerring used Huckin's call as a point of departure for critically examining globalization:

> The globalization of professional communication as a service changes the circumstances (i.e., where, how, under what conditions, by whom, with whom, and for whom) under which the service will be produced. As many of these economic and political changes are accompanied by highly contested ideological agendas, professional communicators increasingly encounter these ideologically charged discourses of globalization in their workplaces and communities. (p. 469)

She then reiterated Huckin's (2002) observation that "the implications of globalization for professional communication programs have remained largely unexplored" (Starke-Meyerring, 2005, p. 469). Recognizing the multilayered meaning of the term, Starke-Meyerring identifies four themes in globalization discourse for examining the work and literacies of technical communicators in the global context:

- the facilitation of globalization through digital networks;
- the resultant pluralized identities and blurred boundaries;
- the increased interactions between diverse local and global discourses; and
- the ideological contestation surrounding globalization. (p. 469)

In the same year, Jeff Grabill (2005) took a similar yet more personal and reflective stance. After acknowledging the origin of his reflection—"My paper and thinking begins with Thomas Huckin's observation at the 2002 Council for Programs in Technical and Scientific Communication meeting that globalization has significant implications for technical and professional communication"—Grabill said,

> I was a member of Huckin's audience in 2002, and I was then—and to an extent still am today—guilty of a lack of global sociopolitical consciousness. Globalization has not been germane to my own research trajectory, a path that has focused on local communities in the US and their literate practices and uses of advanced information technologies. What has changed significantly for me is the move from a teacher of technical communication to a program administrator, and related to that, a move to an institution that takes international issues quite seriously. (p. 373)

Although Grabill's reflection explicitly refers to the persistence of a void in the research on globalization, it also offers an insightful perspective on how context (i.e., moving from a teaching position to an administrative position) impacts the awareness about globalization. In this respect, while Grabill's statement reinforces the idea that globalization has a ubiquitous presence in the administrative and programmatic landscape, it simultaneously shows that, even at the beginning of the new millennium, it is possible to remain sheltered from globalization and to maintain a separation between "local issues" and "global issues."

Despite their differences, the shared calls of Huckin (2002), Starke-Meyerring (2007), and Grabill (2005) carry important lessons regarding the programmatic weight of the global context, and more specifically, its significance for defining the mission of technical communication programs. Indeed, their statements are strong reminders that, if globalization enacts pragmatic responses to shape programs that provide students with the skills to successfully transition into a more diverse workplace, it is also a catalyst for spurring productive debates and reflections that involve the very core and meaning of technical communication's practices. All three emphasize, in different ways, how the global context provides a locus for engaging in critical reflections:

- There are many reasons . . . for TPC scholars and teachers to go beyond our traditional, US-focused instrumentalism and include more international topics viewed from a more critical stance (Huckin, 2002, p. 8).
- As a phenomenon that is highly ideologically contested, globalization calls for a renewed emphasis on critical literacies in professional communication programs (Starke-Meyerring, 2008, p. 470).
- It seems to me that broadly educated people capable of thinking critically about complex sociopolitical issues are the types of people we should be producing in technical and professional writing programs in the 21st century. Only these habits of mind will allow students the intellectual autonomy to

> make good personal decisions and become "good" writers for global organizations of all kinds—organizations that localize products thoughtfully; that design diversity into systems that create safe, entertaining, and beautiful systems and documents; and that treat people and the environment justly (Grabill, 2005, p. 376).

Their shared views on the necessity to approach globalization critically by exploring its deeper ramifications and its "multi-layered meaning" (Starke-Meyerring, 2008, p. 469) open myriad research possibilities. Indeed, the contested nature of globalization—particularly with regard to its ideological, social, and political consequences—raises questions crucial for defining the field as well as the mission of technical communication programs. For instance, and without any pretense of exhaustivity, inscribing a program in the global context fuels interrogations regarding the ways the field can or should position itself vis-à-vis globalization and its polarizing dimension; the ethical issues underlying the globalization of the world (such as environmental questions, human rights, division of labor and labor conditions, and so on); the position of the field with regard to the role of technology and the ways an increasing networked and participatory environment impacts the role of today's technical communicator; and the meaning of a linguistically and culturally diverse workplace for the very definition of communication practices and audience analyses.

Although each interrogation at the core of Huckin (2002), Grabill (2005), and Starke-Meyerring's (2007, 2008) articles could easily constitute the point of departure of a separate study, my main goal in listing examples of questions spurred by the global context is to emphasize the affordances of globalization for sophisticating our definition of the field and of the mission of technical communication programs. In other words, globalization provides a constructive context for reflecting on the identity of the field and on the meaning of its practices. Indeed, all these questions, because they directly engage issues regarding the social and civic responsibilities of technical communicators, echo the positions of a number of scholars who, in the 1980s and 1990s, foregrounded the ideological, political, and social dimension of technical communication and hence promoted an anti-instrumental stance (Herndl, 1993; Katz, 1992; Miller, 1989; Savage, 1996; Sullivan, 1990). In his address, Huckin explicitly acknowledged the parallelism between his stance on globalization in technical communication and earlier works in the field. Referring to Miller's 1989 article, "What's Practical About Technical Writing?" and Dale Sullivan's 1990 article "Political-Ethical Implications of Defining Technical Communication as a Practice," Huckin noted,

> What I am proposing in this paper is not entirely novel. Carolyn Miller, for example, long ago championed the idea that technical and professional communication courses should include attention to civic issues. . . . Not long after, Dale Sullivan added Miller's thoughts by noting that if technical

> communication is seen as praxis, we should expand its scope to include political discourse. (p. 1)

In a way, "listening" to the discussions globalization has enacted in the last 10 years calls for an expansion of our diachronic lens by emphasizing the connections existing between the challenges of the 21st century workplace and the pivotal issues in the history of the field. Thus, because it foregrounds questions that directly speak to the binary opposition between an instrumental and anti-instrumental approach to technical communication, I assert that the global context with, and even because of, all its challenges becomes a means for initiating productive discussions about the disciplinary identity of the field and the mission of technical communication programs.

In this sense, program administrators facing 21st century workplace challenges have much to gain from adopting a stance that considers the synchronic exigencies (*hic* and *nunc*) in light of a more encompassing diachronic perspective. Not only is this a way to better gauge the programmatic significance of the global context, but it also allows for the reconciliation between the current exigencies of the workplace and one of the field's pivotal challenges—the balance between pragmatic needs and humanistic concerns (Dubinsky, 2004).

## CONCLUSION

In this chapter, I have shown how the current context provides a kairotic moment for technical communication program administrators to engage in a diachronic exploration to better gauge the programmatic weight of the global context. By intersecting the discursive and referential manifestations with questions the global context has spurred during these last 10 years, I have emphasized its affordances for sophisticating the mission of technical communication programs. Although this chapter does not claim to solve all the challenges that administrators are facing today, I hope that my contribution offers productive ways for reflecting on the current exigencies and their impact on programmatic planning. I also hope that my contribution provides constructive arguments for justifying the values inherent in positioning programs in the global context. At a time like this, when the pressures to answer the exigencies of the 21st century workplace are so intense and concomitant with the development of skeptical voices, "observing from afar"—or, in Aristotelian terms, to "theorein"[9]—the challenges of the *hic* and *nunc* can enable us to better understand what globalization is, can be, and could be.

[9] As Sharon Crowley (2006) underscored, in "Aristotle's Greek, theorein literally means 'to observe from afar'; it refers to someone sitting in the top most row of the theater." (p. 27)

## REFERENCES


Andrews, D. C. (Ed.). (1996). *International dimensions of technical communication.* Arlington, VA: Society for Technical Communication.

Beamer, L. (1992). Learning intercultural communication competence. *The Journal of Business Communication, 29*(3), 285–303.

Crowley, S. (2006). *Toward a civil discourse: Rhetoric and fundamentalism.* Pittsburgh, PA: University of Pittsburgh Press.

Council for Programs in Technical and Scientific Communication (CPTSC). (2008, May). *Announcing the CPTSC 2008 Graduate Student Diversity, Sponsored by Bedford/St. Martin's* [Newsletter]. Retrieved from http://cptsc.org/newsletters/s08news.pdf

Dubinsky, J. M. (Ed.). (2004). *Teaching technical communication: Critical issues for the classroom.* Boston, MA: Bedford/St. Martin's.

Friedman, T. L. (2005). *The world is flat: A brief history of the twenty-first century.* New York, NY: Picador.

Grabill, J. (2005). Globalization and the internationalization of technical communication programs: Issues for program design. *IEEE International Professional Communication Conference Proceedings,* 373–378.

Herndl, C. G. (1993). Teaching discourse and reproducing culture: A critique of research and pedagogy in professional and non-academic writing. *College Composition and Communication, 44*(1), 349–363.

Hoft, N. (1995). *International technical communication: How to export information about high technology.* New York, NY: John Wiley & Sons.

Huckin, T. (2002). Globalization and critical consciousness in technical and professional communication. *Proceedings of the 29th Annual Meeting of the Council for Programs in Technical and Scientific Communication* (pp. 1–13). Ann Blakeslee (Ed.). Council for Programs in Technical and Scientific Communication.

Katz, S. B. (1992). The ethic of expediency: Classical rhetoric, technology, and the holocaust. *College English, 54,* 255–275.

Lovitt, C. R., & Goswami, D. (1999). *Exploring the rhetoric of international professional communication: An agenda for teachers and researchers.* Amityville, NY: Baywood.

Matveeva, N. (2006). *Teaching intercultural communication in a service technical writing course: Alternative ways of presenting intercultural issues in technical writing textbooks and in real classrooms.* Doctoral dissertation, Texas Tech University, Lubbock.

Maylath, B., & Grabill, J. (2009). The Council for Programs in Technical and Scientific Communication at 35 years: A sequel and perspective. *Programmatic Perspectives, 1*(1), 29–44.

Meloncon, L. (2012). Current overview of academic certificates in technical and professional communication in the United States. *Technical Communication, 59*(3), 207–222.

Miller, C. R. (1979). A humanistic rationale for technical writing. *College English, 40,* 610–617.

Miller, C. R. (1989). What's practical about technical writing? In B. E. Fearing & W. Keats Sparrow (Eds.), *Technical writing: Theory and practice* (pp. 14–24). New York, NY: MLA.

Montgomery, M. A. (2011, May 5). America's international-education programs need a 21st-century makeover. *The Chronicle of Higher Education.* Retrieved from http://chronicle.com/article/Americas/127413/

Nugent, J. (2009). *United States certificate programs in technical communication: A feminist-sophistic investigation.* Doctoral dissertation. Michigan Technological University, Houghton.

Nugent, J. (2013). A survey of U.S. certificate programs in technical communication. *Programmatic Perspectives, 5*(1), 59–85.

Redden, E. (2011, June 3). The end of internationalization? *Inside HigherEd.* Retrieved from http://www.insidehighered.com/news/2011/06/03/international_educators_debate_the_why_behind_their_profession

Savage, G. J. (1996). Redefining the responsibilities of teachers and the social position of the technical communicator. *Technical Communication Quarterly, 5*(3), 309–327.

Schneider, A. I. (2011, June 1). International-education programs should be continued. *The Chronicle of Higher Education.* Retrieved from http://chronicle.com/article/article-content/127734

Spring, J. (2009). *Globalization and education: An introduction.* New York, NY: Routledge.

Starke-Meyerring, D., Duin, A. H., & Palvetzian. T. (2007). Global partnerships: Positioning technical communication programs in the context of globalization. *Technical Communication Quarterly, 16*(2), 139–174.

Starke-Meyerring, D., & Wilson, M. (Eds.). (2008). *Designing globally networked learning environments: Visionary partnerships, policies, and pedagogies.* Rotterdam/Taipei: Sense.

Sullivan, D. (1990). Political-ethical implications of defining technical communication as a practice. *Journal of Advanced Composition, 10,* 375–386.

Thrush, E. (1997). Multicultural issues in technical communication. Foundations for teaching technical communication. In K. Staples & C. Ornatowski (Eds.), *Foundations for teaching technical communication: Theory, practice, and program design* (pp. 161–178). Greenwich, CT. Ablex.

Weiss, T. (1995). Translation in a borderless world. *Technical Communication Quarterly, 4*(4), 407–423.

CHAPTER 9

# Users, Not Solutions, First: Problem-Solving for Program Administrators

*M. Ann Brady and Karla Saari Kitalong*

Program administrators are problem-solvers. Sometimes the problems are conceptual, such as deciding on the theoretical dispositions that infuse a program. Generating learning goals compatible with those dispositions requires administrators to initiate curricular reviews and innovations as well as to design and carry out assessment to reinforce potentially successive gains. And if assessment is to support instructors' creative, hard work, then administrators need to turn assessment results into instructor enrichment initiatives, such as pedagogical workshops, seminars, and discussions. Many program administrators are also charged with the more mundane tasks of keeping a program running: managing budgets, evaluating instructor performance, and keeping track of material resources.

The role of the problem-solving program director extends to both creative and mundane tasks and is among the professional activities characterized by Schön (1982) and Mirel (2004) as unprecedentedly complex: "Professionals are called upon to perform tasks for which they have not been educated" (Schön, 1982, p. 15), tasks imbued with "uncertainty, disorder, and indeterminacy" (Schön, 1982, p. 16), highly contextualized, and "tied to the interests of diverse stakeholders" (Mirel, 2004, p. 22). Moreover, "many possible and equally legitimate moves and strategies" can be taken to accomplish a workable solution (Mirel, 2004, p. 16). Because few cut-and-dried responses exist, professionals need to develop multiple strategies for "functioning in situations of indeterminacy and value conflict" (Schön, 1982, p. 17) and devise their own ways to combine strategies in response to complex situations.

Given our combined nearly 20 years of experience as both technical communication and composition program directors, we have addressed, if not solved, most all the problems in this book. From designing programs to stretching budgets, we have found the work of problem-solving to be both challenging and satisfying, if anything but linear and straightforward. Yet current frameworks for problem-solving found in management textbooks are based on technical rationality in which "professional activity consists in instrumental problem solving made rigorous by the application of scientific theory and technologies" (Schön, 1982, p. 21). Such a problem-solving approach suggests a straightforward path: define the problem, design the solution, test the solution, implement it, and then evaluate it. Simple steps such as these are often embellished in industry to increase production and profitability, with sprints and scrums keeping designers on task (Deemer & Benefield, 2007). By contrast, as our introductory paragraphs suggest, problem-solving for program administrators cannot be described linearly. We must define and contextualize problems, involve as many stakeholders as possible, revisit our assumptions and processes—often many times until we're satisfied we have clearly articulated the problem and considered all potential solutions—and remember always to go slowly. The bottom line, as Schön (1982) put it, is that "complexity, instability, and uncertainty are not resolved or removed by applying specialized knowledge to well-defined tasks. If anything, the effective use of specialized knowledge depends on a prior restructuring of situations that are complex and uncertain" (p. 19). In short, too much specialized knowledge applied to complex situations can relegate the stakeholders to a technically rational response and therefore predispose a "solution first" disposition.

We model our problem-solving approach on what Scandinavian computer systems designer Pelle Ehn (1988) termed *work-oriented design*. Growing out of his commitment to "social and technical change," he advocates a shift in the design process to include "ordinary people at their workplace" (p. 7). Although his goals focus on computer technology and advance "industrial democracy" (p. 4), we see clear connections between Ehn's work and ours. Just as computer design benefits from including working people early in the planning process, so too can technical communication programs engaged in solving complex problems benefit from faculty and student perspectives about their needs and interests. Our resolve to include these stakeholders is also informed by interdisciplinary technical communication scholarship that has refined Ehn's work, resulting in user-centered design principles that we see being applied widely in workplace and academic settings today (Carroll, 2011; Johnson, 1998; Nielsen, 1993; Norman, 1998). What links these user-centered theories is a commitment to exploring problems with stakeholders likely to be affected by them. We believe that these approaches share a participatory approach to problem-solving that "supports, reinforces, and furthers our collective efforts to place users at the center" (Brady & Hayenga, 2012). It enables inclusion and fair representation of the voices of multiple stakeholders in ways that accommodate diversity and address conflicting goals and agendas (Anderson, 1995). The Department of Humanities at Michigan Technological University recently engaged in such a participatory process to redesign a departmental computer lab after trying two other problem-solving approaches: emergent and solution first. Lengthy and decidedly nonlinear, perhaps even rhizomatic, the participatory process allowed us to reenvision our long-standing computing facility. In this chapter, we narrate and critically reflect on the story of this complex redesign process.

## SNAPSHOTS OF THE CENTER FOR COMPUTER-ASSISTED LANGUAGE INSTRUCTION

The Center for Computer-Assisted Language Instruction (CCLI) in Michigan Tech's Department of Humanities was envisioned first by Cynthia and Richard Selfe in the early 1980s and opened its doors in 1985. Over 25 years, the CCLI garnered national recognition as an innovative solution to the problem of encouraging students and faculty from nontechnical disciplines such as English and the Humanities to use computers in their learning and teaching. One of the many reasons for the Center's success and longevity is that the Selfes encouraged "emergent" problem-solving. The premise of such a process—that a collective solution represents a better solution—requires diverse stakeholders, who can contribute missing pieces of information critical to solving a problem that all are experiencing. Important components of emergent problem-solving are situatedness and reflection, as well as the way the solution emerges iteratively as people, contexts, artifacts, and bureaucracies interact. As such, it is an effective application of user-centered design: as solutions emerge, participants share them with others, discuss them, try them out, and discard them until the collective agrees on the most optimal solution. Emergent problem-solving, in the context of our discussion, also depends on stakeholders being familiar with the context in which the problems arise and familiar enough with each other to discuss problems spontaneously and candidly. Because the Selfes used the lab themselves, they were frequently aware of problems as they surfaced. And because they were in daily contact with CCLI users, they could observe how problems impacted productivity, subsequently initiating discussions with instructors and students about possible solutions. Emergent problem-solving, thus, became standard practice in the CCLI and was serving most users very well when Karla Kitalong worked there, as she describes in the next section.

### Karla Remembers: 1992–1999, 2007

I worked in the Humanities Department from 1992 to 1999 as a lecturer with specific responsibilities for faculty computer support and undergraduate academic advising. The CCLI was, at the time, a model facility for Humanities computing. Back in 1985, Cynthia and Richard Selfe had envisioned the space as a participative, writing-centered environment—a "cyborg environment . . . made up by both people and machines" (Selfe, 2005, p. 56). The people, including students, teachers, and lab consultants, were considered to be at the center of this space. The machines—"seldom entirely clean, always well used, and constantly in need of upgrading"—were also crucial, as was the space itself, the arrangement of which "encourage[d] some activities and discourage[d] others" (Selfe, 2005, p. 56). The user-centeredness was reaffirmed in the everyday ambience of the lab. Art adorned the walls, and the consultants on duty wore colorful kimonos. Toys, including stuffed animals and Nerf dart guns, were scattered about, ready for creative play. Departing from the best practices of the time, the lab was equipped with a refrigerator and microwave, and eating and drinking were encouraged, the rationale being that writing is a sustained, labor-intensive activity that requires sustenance, and should someone spill soda on a keyboard, it would be inexpensive to replace.

The CCLI was framed by two sets of goals—instructional goals and operational goals—that had been established by interested faculty when the facility was in its planning stages (see below). Although the goals were subject to change based on the input of the stakeholders (Selfe, 2005, p. 64), they remained substantively unchanged for some 20 years. The user-centered articulation of instructional goals, or "shared pedagogical values," served several purposes for the CCLI.[1] When the faculty participate in the development of pedagogical goals and feel empowered to suggest changes to those goals, they are more likely to be invested in the facility's success. The shared goals also help create an "environment that mirrors the instructional goals a faculty considers important and students will find useful," thereby "encouraging broad support" within the department and the institution. Finally, student stakeholders benefit "when faculty have listened to what students have said" about the challenges associated with literacy learning (Selfe, 2005, p. 64). In short, the goals afforded mutual reinforcement of teacher and student interests and situated the space in the larger institutional and disciplinary context.

## CCLI Instructional Goals

- Goal #1: Encourage students to practice communicating (through text, graphics, sound, still and moving images) as often as possible and to improve their skill as communicators.
- Goal #2: Support the concept of communicating for a variety of purposes, using a variety of communication strategies, aiming for a variety of audiences.

[1] See http://web.archive.org/web/19990210072201/http://www.hu.mtu.edu/ccli/

- Goal #3: Promote the understanding of communication as a complex, socially situated, and political set of processes through which humans act and make meaning.
- Goal #4: Help students recognize that communication takes place within, is shaped by, and serves to shape both social and political contexts.
- Goal #5: Encourage collaborative exchanges among communicators and teachers of communication, using strategies such as peer feedback sessions, student-teacher conferences, and student-tutor conferences.
- Goal #6: Support effective process-based instruction in communication for students and teachers who request it using strategies such as communication-intensive classes, conferences, and individual practice with communication.
- Goal #7: Help communicators learn to be critical readers of their own and others' communication.
- Goal #8: Encourage communicators to learn and share successful communication strategies.

## CCLI Operational Goals

- Goal #1: Purchase and maintain software and hardware that will support a socially situated, process-based approach to communication and the teaching of communication.
- Goal #2: Provide CCLI access to any student enrolled in a communication-intensive class and to any instructor teaching such a class.
- Goal #3: Make sure the CCLI is flexible enough to support communication communities of all sizes individuals, small groups, whole classes, etc.
- Goal #4: Make sure the CCLI is administered on a policy level by a communication faculty member and staffed on a daily basis by consultants who have expertise both in communicating and the teaching of communication.
- Goal #5: Provide hours that are flexible enough to accommodate individuals' communication habits and needs as well as groups' communication habits and needs.
- Goal #6: Provide adequate technical support for maintaining, repairing, and modifying computer hardware and software so that communication faculty do not have to assume this role.
- Goal #7: Ensure that CCLI policies encourage process-based communication, communication as thinking, and communication as social action.

Although the pedagogical and operational goals remained largely unchanged for the lab's 25-year history, a longer and much more fluid procedure document appears to have evolved as new capabilities were introduced and user expectations adjusted in response. This document would have been the product of an ongoing emergent problem-solving process, a situated, reflective, and iterative process that brings people, contexts, artifacts, and bureaucracies together to generate the best possible solution for all. For example, as the network grew in importance during the 1990s, a whole section on communication was added to the policies and procedures document through what was undeniably an emergent problem-solving process. Some of these policies (see the below 10 rules) seem incongruous from today's vantage point, but they were responsive to the immediate context.

## Communications

- Avoid frivolous electronic mail or messages.
- Do not transmit annoying, harassing, or unsolicited material to other users.
- Do not use e-mail for commercial advertisement or personal gain.
- Using e-mail for any form of advertisement could easily be construed as "unsolicited mail" and should be avoided. This includes advertising for University groups, no matter how innocent your intentions may be.
- Use mailing lists, news groups, and aliases for their intended purposes.
- Forging computer mail is a very serious offense and strictly prohibited. E-mail makes no pretense at being very secure.
- DO NOT use the network or computer equipment for illegal activity (e.g., piracy, computer fraud, etc.). People doing so may face disciplinary action or criminal prosecution.
- There are no restrictions on allowable language which users of the CCLI may use when communicating via computer, but good taste and consideration of others should be used at all times. If you consider your audience carefully before using computer communications, you should encounter no troubles.
- Check your e-mail at least once a week. Important notices are sent to users via e-mail.
- It is highly recommended that appropriate subject lines are used as a courtesy to others.

To map these policies to an emergent problem-solving process would take a much more in-depth analysis than is possible here, but the emergent nature of the communication policies can be inferred. For example, the presence of emotionally laden language, including words like *frivolous*, *annoying*, *illegal*, *good taste*, and *unsolicited*, hint at the exigencies that might have triggered the inclusion of these rules in the document. I can imagine users of the lab as busy as today's Internet users, but whose Internet connections were much slower than those to which we have become accustomed. I also remember well a time when e-mail had not yet become second nature; for instance, Rule 9 asks that users check their e-mail at least once a week. Surely this rule emerged from a situation in which a number of the lab's users did not know when changes occurred in the lab because they didn't read their e-mail. Clearly, if these rules were revisited today, some would no longer be necessary though others would certainly persist. In fact, Rule 9 might well have disappeared for a time because although most people checked e-mail regularly when it was the best (if not the only) mode of Internet communication available, today's students have largely moved beyond e-mail to communicating via text or instant messaging and social networking sites (Gussow, 2005). Emergent problem-solving would ensure that the collective was involved in evaluating and amending the rules of the road as the CCLI and the world around it changed.

"This is a connected place," Selfe wrote (2005, p. 57), and indeed, even in the early, pre-Internet years of the lab, a local area network enabled collaborative writing, class discussions, and hypertextual reading (Cooper & Selfe, 1990). "On good days," Selfe (2005) wrote, "these humans and technical systems interact productively, together, constituting a wonderful synergistic dance" often "invisible" to those who use the facility and even sometimes to those who administer it (p. 57). The 10 rules that had been instituted emergently to govern network communication were obviously meant to make good days out of bad ones. Reading between the lines, the policies concerning access to the network allude to the influence students exerted in making the rules in the CCLI. Gaming was permitted, if not overtly encouraged. The playful atmosphere of the lab was retained, but the rules were written to teach what courtesy looks like in electronic environments; in fact, several guidelines look like adaptations of the Golden Rule. In keeping with the instructional goals that had been solidified in the mid-1980s, the language of the communication policies emphasizes diversity—acknowledging that different people may have different standards of "allowable language," for example—but overall, the rules convey an expectation that the lab's users will be serious, purposeful, and considerate; will be conscientious stewards of the lab's resources; and will practice good rhetorical principles, particularly audience awareness.

At the time Selfe (2005) was writing his book, the "most successful computer-supported environments for English and language arts" could boast a "20-year history of supporting" the work of "programs, departments, and students" through a commitment to user-centered design: "supporting the changing needs of a wide range of students and the changing instructional goals of a wide range of faculty"; in short, the "process of their design—and continual reassessment and redesign . . . involves those people who have a stake in their operation" (p. 59). Selfe was referring to the CCLI's history of involving interested stakeholders, including students, faculty, staff, and even certain members of the university community. Indeed, teachers, primarily those who taught media-intensive classes, worked extensively in the lab.

Nonetheless, by the mid-1990s, students were the acknowledged primary users of the space. They provided the day-to-day labor, often on a volunteer basis; they planned, carried out, and assessed all behind-the-scenes maintenance; they lobbied for sophisticated software, high-end computers, and color printing, and then took it upon themselves to learn how to use the new additions so that they could assist other users. And they negotiated the rules. As this process developed and accelerated, faculty interests fell increasingly by the wayside. That's why, when I was hired to the department in 1992, my initial job description involved bringing faculty offices online, helping to find funding for faculty computer upgrades, and teaching faculty to use the computers they found on their desks, which were often hand-me-downs from lab upgrades. My position was intended to bridge a growing divide between faculty and student computing that perhaps began when the university shifted from a centralized to a distributed technology funding model, the core of which was the institution of rather steep student lab fees. The fees allowed departments to equip state-of-the-art student labs, but the rules prohibited the use of lab fees in support of faculty computing. In some ways, then, despite a dedicated technical support staff member, Humanities Department faculty office computing capabilities continued to lag behind what was available to students in the CCLI.

I finished my doctorate and left Michigan Tech in 1999; at that time, the CCLI was still a thriving enterprise with a committed community of users—students, primarily, but a few faculty members and administrators as well—who both depended upon and were invested in its success. After an 8-year absence, I interviewed for a faculty position at Michigan Tech in December 2007. By this time, the Selfes and several other key faculty members had moved on, leaving behind a very different department. The interview process included a tour of the lab that left me nostalgic. Despite a recent fresh coat of colorful paint, the CCLI had changed little in the intervening years—same décor, same pod configuration, same tables and chairs. But it had become a ghost town, or perhaps more accurately, a shrine. The artwork, toys, and other artifacts contributed by the once-thriving community were still there. A California Raisin puppet had once served an important function: suspended from a doorknob, it had contained a wandering toddler who was inexplicably freaked out by it. Now it hung lifeless, purposeless, from a shelf bracket. A pen and ink cartoon drawn by a long-gone Scientific and Technical Communication major had once depicted an iconic moment in the lab. Now decontextualized, its exigency forgotten, it still occupied a place of honor on the north wall. The beautifully framed circuit boards, meant to render visible and thereby demystify the inner workings of the lab's computers, were still prominently displayed but somehow out of place. The platform-based split—Macs in one room, PCs in the other—persisted despite the fact that textual, visual, and aural technologies now functioned seamlessly on both platforms.

But, as Ann will now explain, the surface-level sameness belied the inevitable curricular and disciplinary innovations that new constituencies introduced.

## Ann Remembers: 2003–2009

When I took a faculty position at Michigan Tech in 2003, the CCLI was a space still known to "facilitate playful interaction as well as serious labor" (Bridgeford, Kitalong, & Selfe, 2004, p. 3). It did so, however, for a particular group of students—those in design and media. The CCLI fostered a dynamic production process among these students and encouraged them to engage in active reflection. Instructors in these areas designated their classes "computer intensive," thus placing the CCLI and its technologies at the center of their pedagogy and underscoring the importance of using its technologies to engage in the creative problem-solving their assignments required. These faculty often met their classes in the CCLI, a place where they could bring students together for brief lectures or reviews before opening up the remaining time to individualized instruction or group work. With its clustered computer tables and informal atmosphere, the lab served as a think tank for student projects and a launching site for their products, including videos, websites, and print documents. An annual showcase was often located in the CCLI and in the classrooms close by, an event that welcomed the larger community to view the lab as central to innovative student work in media and design. In these instances, design and media instructors and student users benefited from the CCLI.

Those in other areas of the department generally did not. The CCLI served a department of humanities, composed not only of media and design, but also of technical communication, composition, modern languages, philosophy, sociolinguistics, English, education, and communication and culture studies. Faculty in all areas of the department were invited to designate their courses "computer intensive" and thus to make the CCLI and its resources available to students. But most did not, and for several reasons. Most courses did not have a "computer intensive" history, and thus no community lore existed about how to make use of a computer lab when teaching a course outside media and design. Such lore communities might have been built with orientation programs for novices or ongoing training in how to make practical use of the space. Without such support systems in place, however, new and returning faculty often overlooked the CCLI in what were otherwise innovative course designs. When I arrived on the Tech campus, then, the CCLI served a limited user group, its many resources untapped by the majority of faculty and students.

Over the next years, several changes, both in departmental and student cultures, compounded the problem of limited use that the CCLI faced when I arrived. Although superficially unrelated, these changes undermined the emergent problem-solving process that had contributed to the lab's success, transforming it into a solution first approach. More specifically, they foreclosed opportunities to discuss problems as they arose and narrowed stakeholder diversity, thus limiting the possibilities for robust solutions.

The first change was the Selfes' departure, which left a gap in the CCLI's emergent problem-solving practices. Although emergent problem-solving is a collective process, research indicates that it benefits from the presence of "decisive and reasonable" participants. That is, if there are too many random solutions or too many indecisive problem-solvers "then the collective fails to be better than the average individual" (Johnson, 1999, p. 35). Although all participants were invested in identifying solutions to the various problems that arose in the CCLI, students came and went from one year to the next, though instructors' needs changed semester by semester and course to course. The Selfes' long-term commitment to the lab's success positioned them at the center of the collective and identified them as the most consistent "decisive and reasonable" problem-solvers in it. Without their presence, the collective lost key players in emergent problem-solving, a gap that subsequent interim directors could not fill.

Well-qualified and collegial though they were, with limited 2-year appointments, interim directors had just enough time to get to know faculty, but not enough to get to know their needs as users (or potential users) of the lab. Lacking knowledge of the CCLI's emergent problem-solving history and without time to explore a vision for its future, interim directors maintained the lab as it had been—the same computer configurations, the same spatial design, the same informal ambience, the same solutions. What had been fresh and dynamic answers to emerging problems became codified and stale. Emergent solutions became static defaults that put them, not the user, first. So for instance, new PCs were purchased every 3 years because purchasing had always been scheduled that way, not necessarily because users needed new technology then or because they had needed it long before the 3 years had elapsed (E. Smith, personal communication, May 13). "Solution-first" problem-solving (Carroll, 2011) is particularly compelling in the uncertain circumstances that the CCLI was experiencing. With no obvious "decisive and reasonable" problem-solvers at the center of the CCLI collective, solution-first was expedient. It required no communication among stakeholders; it demanded no shared vision of problems and opportunities; it functioned without a context since the active faculty users of the lab had departed, taking the details of the institutional memory with them. In short, it offered concrete (and inflexible) solutions. Over the next 4 years, those associated with the CCLI, either as administrators or faculty, would use solution-first problem-solving consistently, much to the detriment of new users with new problems and the lab as a dynamic learning space.

The second change—ironic in light of the diminished use that I have described—was one program's increasing interest in and use of the CCLI: Scientific and Technical Communication (STC). When I assumed responsibilities as STC director, the program had always required its students to take courses in both media and writing. Although the media courses that STC students took were regularly taught in the CCLI by faculty specializing in this area, none of the writing courses were taught there. Aware of the possibilities for collaborative writing, writing assignments that addressed design problems, and peer review in the CCLI, I identified one "writing" course that could serve as a pilot. This cross-disciplinary and upper-division course, an introduction to the practices, theories, and genres of technical communication, was an STC core requirement as well as one possible required elective for students in the sciences, engineering, and business. It offered instruction in collaborative design and writing, professional editing, and usability testing, all of which could be supported by CCLI technologies. With multiple sections of this course, capped at 25 students, all enrolled to capacity each semester, several problems arose.

The most revealing was one of scheduling. Media and design classes now found themselves competing for CCLI classroom time with those in technical communication. Interim directors sought alternatives, but these were based on outdated information about who used the CCLI. Important questions were left unasked: Who are the new users? What are their needs? And how do those needs intersect or compete with those of established user groups? Without this information, solution-first remedies—for instance, dividing the lab space, half to media classes and half to writing—served no one very well. At times, media and writing classes were cross-booked for the same space; at others, two writing classes were booked for the same time. The result was an overcrowded CCLI during regularly scheduled classes and a virtually empty CCLI at other times.

Rapidly diversifying users whose needs were left unaddressed presented problems exacerbated by a third change: increased laptop use. In its heyday, the CCLI had offered a comfortable space for design teams to gather, brainstorm, problem-solve, collaborate, and produce. It had offered a cutting-edge spatial design, replacing rows of computer desks, all facing the front of the classroom, with clustered computer tables where students could see each other's screens and swivel this way or that to consult, face to face, with each other. Here, unlike in other labs on campus (and indeed nationally), the affective component of cognition was acknowledged: people felt relaxed and comfortable there, so they were more inclined to engage in creative problem-solving (Norman, 2004). The CCLI's walls were adorned with posters and paintings; a refrigerator and microwave encouraged users to bring food and beverages into the lab; all these attributes contributed to a welcoming workspace. With increasing numbers of students, both in Humanities and the Sciences and Engineering, using laptops, however, such benefits were no longer unique. No longer did students need to go to the CCLI to find an atmosphere conducive to creative focus; they could work from anywhere on Tech's networked campus or from their own homes. And no longer did collaboration require that they meet in the CCLI where groups could gather around a computer screen, because they could haul their laptops anywhere or collaborate electronically.

The problem of a changing student culture was addressed, once again, by putting a solution in place without considering user needs. Because communication about problems had been spontaneous and informal under the emergent problem-solving model, no formal communication channels had been established. Interim directors, operating as solution-first problem-solvers, assumed that informal communication would work just as it had in the past. What they did not take into account, however, was that a CCLI collective—users with a shared context focusing on a problem that all experienced—no longer existed. Media and design students were no longer the primary users of the lab, and new users from the Humanities and Sciences and Engineering could have benefitted from a different form of communication: formal channels that would put them in touch with lab administrators and that would allow them to recommend ways that the CCLI might better serve their needs as cross-disciplinary laptop users. Solution-first problem-solving remained the order of the day, however. No formal channels were established, and CCLI use continued to plummet.

## Transition from CCLI to Humanities Digital Media Zone: 2009–2011

In 2009, the department appointed a new CCLI Director, Dr. Erin Smith, who had taught in the department for a number of years, and as such was fully aware that the CCLI no longer served students or faculty well. Consulting with faculty and program directors in technical communication, composition, modern languages, communication and culture studies, and visual and performing arts, Smith immediately began planning for a renovation in 2009. Returning to a user-centered design tradition, but now with a participatory problem-solving approach, Smith and an active planning committee sought input from department faculty and students as well as external stakeholders, including university physical plant engineers, networking and information technology staff members, and an outside architectural firm.

Throughout the process, Smith and the committee remained committed to ensuring that they, the lab staff, and student employees would support faculty members' techno-pedagogical efforts. The result is a completely new space, with a new name: the Humanities Digital Media Zone (HDMZ), a multidisciplinary, technology-rich resource center. Designed to create a flexible and sophisticated technological work and social environment, the HDMZ serves a broad range of Humanities-centered educational goals and technology practices. It represents the consolidation of the former CCLI and the Modern Language Lab, as well as a classroom and a number of other discrete spaces. It includes spaces that address the needs of a diverse group of users: a laptop classroom; a digital media studio; two recording and listening rooms; a resource desk for modern languages and media development; a multipurpose space dedicated to a range of activities and research; administrative, staff, and student organization offices; and a commons area for work on laptops and other mobile devices. Resources available to students include digital SLR and video cameras, voice recorders, smart pens, graphics tablets, video recorders, and other language and media resources.

The technological resources are important, of course, but what distinguishes the HDMZ from the CCLI as it had functioned by the time I arrived is the use of a participatory problem-solving process. In an interview with Dr. Smith, we learned that, for her, "Solutions were not first; questions and problems were first," in her planning. "Listening," Smith said, "reveals the problems and the needs." In discussions with faculty interest groups, Smith thus asked open-ended questions, such as, "Why is it important for you to be in this space? Being here, what do you want and what do you need?" Well aware of the iterative nature of participatory design, Smith reported that before she could "grasp the solution," she had to "see the bigger picture and understand the larger stakes." The planning committee thus engaged in a figurative "card sorting" activity, similar to a technique widely used by information architects and other IT professionals: "Card sorting is a . . . method for finding patterns in how users would expect to find content or functionality" (Spencer & Warfel, 2004). It begins with brainstorming—in our case, listing all the possible uses for the lab space, all the courses that were and could be taught there, and all manner of pedagogical needs. A typical follow-up to card sorting is clustering, also known as affinity diagramming, in which related ideas, applications, and technologies are grouped together. Card sorting and clustering remind us that when the problems are complex and engage multiple stakeholders, any number of solutions might be found. This participatory strategy foregrounds all stakeholder needs and encourages designers to keep their options open as long as possible.

Instead of the "solution first" technique of focusing on curricular needs, platform-specific needs, or software needs, we spent time understanding how teachers wanted to teach in the lab. This participatory problem-solving led us to set aside arguments over whether we needed Adobe Connect or Skype, for example, or whether technical communication classes needed RoboHelp or could manage with a shareware online help package. Instead, we discussed what pedagogical requirements were in play, weighed costs and benefits, and sought to achieve solutions that could be shared among a range of constituencies. As a bonus, in addition to improving the services and technology offered to students and faculty, the renovation made visible the interdisciplinary and global nature of the research, teaching, design, writing, and production work done by Humanities faculty and students. With its open floor plan, range of technology resources, and opportunities for interaction and collaboration, the HDMZ strives to situate technology use as a complex human practice undertaken in a global context.

## Plans and Lessons: 2011–

As administrators, we are committed to infusing the programs we direct with user-centered design principles. To do so, we are equally committed to participatory problem-solving, engaging as many stakeholders as possible in the messy process of solving problems that stand in the way of productive teaching, learning, and collaboration. Doing so, we believe, can provide people with a "pleasurable sense of empowerment" (Norman, 2011, p. 10) because it makes programmatic goals understandable, accessible, and memorable. Critically reflecting on Michigan Tech's CCLI to HDMZ history has given us a deeper understanding of three problem-solving strategies that have been used (with varying degrees of success and failure) to integrate the lab with the curriculum and to acknowledge that faculty are the experts concerning their pedagogical needs. We have recounted how two problem-solving strategies—emergent and participatory—have moved us closer to a user-centered design for our technology lab. And we have demonstrated how another—solution-first—is not user-centered at all because it is more concerned with expedient solutions than with understanding the human contexts in which those solutions are deployed.

Participatory problem-solving worked well for the HDMZ redesign because it factored in visibility: "The relationships among the user's intentions, the required actions, and the results are sensible, nonarbitrary, and meaningful" (Norman, 2011, p. 22). Visibility will also be crucial as we start thinking about how to assess people's experiences in the facility. Over the next year, we will begin to ask the following questions: Does the HDMZ adequately meet the needs that the department articulated 2 years ago? Do our internal and external stakeholders have different views on the lab's success? Has the new design created previously unseen needs or desires, and how should these be accommodated? What kind of culture do we want to foster within the HDMZ and by extension, within the department? What should our next steps be?

Because of the deeply contextualized nature of participatory problem-solving as it leads to user-centered design, we do not claim a "Eureka!" moment, a high point of success in finding the ideal way to administer programs. User needs, the contexts in which they work, and the technology that they use will change, and we may find ourselves moving back to an emergent problem-solving approach in the future. We can, however, offer observations about what we have learned as we have observed and now recounted the CCLI to HDMZ transition. We have learned that the problem-solving strategies that administrators use to design their programs, whether explicit or tacit, have profound effects on the users, technologies, and bureaucracies that constitute the programs. In other words, even if they are not made explicit, problem-solving approaches still operate and can advance or undermine programmatic design goals, be they user-centered or not. Given this, as program administrators, we have learned that having a "vision" for a program is not sufficient. We need to ask, along with Dr. Smith,

"How specific should the vision be? How should it be expressed so that it can be accomplished but remain flexible?" Such are the questions that undergird administrative problem-solving, not necessarily because they offer a solution, but because they can engage all users in the problem-solving process.

## REFERENCES


Anderson, P. V. (1995). Evaluating academic technical communication programs: New stakeholders, diverse goals. *Technical Communication, 42,* 628–633.

Brady, A., & Hayenga, E. (2012). Moving on and beyond: Technical communication program assessment at Michigan Tech. *Programmatic Perspectives, 4*(1), 113–135.

Bridgeford, T., Kitalong, K. S., & Selfe, D. (Eds.). (2004). *Innovative approaches to teaching technical communication*. Logan, UT: Utah State University Press.

Carroll, J. (2011, April 1). *Scenarios and design.* Lecture conducted at Michigan Technological University, Houghton, MI.

Cooper, M., & Selfe, C. L. (1990, December). Computer conferences and learning: Authority, resistance, and internally persuasive discourse. *College English, 52,* 847–869.

Deemer, P., & Benefield, G. (2007). *Scrum primer: An introduction to agile project management with Scrum.* Retrieved from http://www.rallydev.com/documents/scrumprimer.pdf

Ehn, P. (1988). *Work-oriented design of computer artifacts.* Stockholm, Sweden: Arbetslivscentrum.

Gussow, D. (2005). *Teens getting more tech savvy, but find email passe*. Washington, DC: McClatchy-Tribune Information Services. Retrieved from http://search.proquest.com/docview/463571377?accountid=28041

Johnson, N. L. (1999). *Collective problem solving: Functionality beyond the individual.* Unpublished paper. Los Alamos, NM: Los Alamos National Laboratory. Retrieved from http://citeseerx.ist.psu.edu/viewdoc/download?doi=10.1.1.74.408&rep=rep1&type=pdf

Johnson, R. R. (1998). *User-centered technology: A rhetorical theory for computers and other mundane artifacts.* Albany, NY: State University of New York Press.

Mirel, B. (2004). *Interaction design for complex problem solving: Developing useful and usable software.* San Francisco, CA: Morgan Kaufmann.

Nielsen, J. (1993). *Usability engineering.* San Francisco, CA: Morgan Kaufmann.

Norman, D. (1998). *The design of everyday things*. New York, NY: Doubleday.

Norman, D. (2004). *Emotional design*. New York, NY: Basic.

Norman, D. (2011). *Living with complexity.* Cambridge, MA: MIT Press.

Schön, D. (1982). *The reflective practitioner.* New York, NY: Basic.

Selfe, R. J. (2005). *Sustainable computer environments: Cultures of support in English studies and language arts.* New York, NY: Hampton.

Spencer, D., & Warfel, T. (2004, April 4). Card sorting: A definitive guide. *Boxes and Arrows*. Retrieved from http://www.boxesandarrows.com/view/card_sorting_a_definitive_guide

CHAPTER 10

# Intersections Between a Technical Communication Program and an Engineering Department

***Julie Dyke Ford***

Since 2004, I have served as Program Director for the Technical Communication program at New Mexico Tech. When I stepped into that position over 8 years ago, I had just completed my first year as an assistant professor of Technical Communication. With a full teaching load and the demands of maintaining a rigorous research agenda on my path toward earning tenure, I was nervous about taking over this role. However, because the other members of the faculty in my department at that time were either already serving in administrative positions or unqualified or unwilling to take the role, I was equally nervous about what would happen if I did *not* take over as Program Director. Our program, which had been established in the early 1980s as one of the first bachelor's of science degrees in technical communication in the country, was suffering from low enrollment and lack of vision. Here was an opportunity to restore the program to the respected status it once had as an applied program with a balanced curriculum that provided students with excellent preparation to enter scientific, technological, and engineering settings.

I'd be dishonest if I claimed that this job has been without challenges. Over the last several years, our program has weathered high turnover of faculty, hiring freezes, dips in enrollment, and limited travel funds and resources. Our program has moved from one building to another; faculty have played musical offices; and our student lab has moved, shrunk, and been reconfigured. The administrative assistant, who was a wealth of knowledge concerning the program's beginnings, history, and day-to-day matters, had retired in 2010, ending her 30-year career working for our program. However, despite these trying circumstances, I am pleased to be able to say that thanks to some amazing colleagues, exceptional students, and my own hard work, I am ending my term as Program Director with our program in good standing. Our enrollment is back to a steady rate, we hired two new Technical Communication tenure-track faculty members who will join our program during the fall of 2012 (one of whom will take over as Program Director), and the last few years have yielded some fantastic opportunities for students, our faculty, and the program as a whole.

After 8 years, it is time for me to step down as Technical Communication Program Director and be afforded more time to devote to research and continue to develop interdisciplinary initiatives, interconnected activities that fulfill shared goals and offer value for collaborating disciplines, on our campus. Although I consider the last 8 years a "good run," I am equally excited about the prospects some of my future work may yield for the program. With my involvement in a Department of Education grant that supports integrating communication into Science, Technology Engineering, and Math (STEM) graduate curricula, I expect continued exposure of our Technical Communication program not only on our campus but also as a model for other institutions. Additionally, in my new role as a jointly appointed associate professor in Technical Communication and Mechanical Engineering, I have identified several opportunities to integrate these two disciplines in ways mutually beneficial.

I have written about various aspects of our program, including our senior thesis requirement, our corporate advisory board, and our peer-reviewed e-journal *Xchanges*, in prior publications (Ford, Bracken, & Wilson, 2009; Ford & Lanier, 2011; Ford & Newmark, 2011; Ford, Newmark, & Durao, 2013 forthcoming; Newmark & Ford, 2013), so to avoid being redundant, I focus this chapter on our program's tie to engineering. This connection to engineering programs at New Mexico Tech has significantly impacted our program's status on campus and affected our plotting of future directions. My narrative includes reflection on the evolution of our collaboration with engineering and is supported by literature confirming that intersections between technical communication and engineering not only make sense but can also be fruitful.

## ENGINEERS, SCIENTISTS, AND, OH YEAH, COMMUNICATORS

Before presenting information about our program's tie to engineering, some initial context about New Mexico Tech is necessary for understanding the positioning of our Technical Communication program within the university. With a total student body (graduate and undergraduate populations combined) of 2,009, New Mexico Tech is a small-school setting where the faculty-student ratio is 11:1. However, unlike many schools of comparable size, New Mexico Tech is a state-supported research university granting bachelor's, master's, and doctoral degrees in scientific and engineering fields. With more than a dozen affiliated research divisions, such as the Petroleum Research Recovery Center, the Energetic Materials Research and Testing Center, and Langmuir Laboratory for Atmospheric Research, New Mexico Tech is known as a center for world-class research in technical and scientific areas.

When I began my position at New Mexico Tech, I was excited about the dominant engineering and science culture and the cutting-edge research projects happening all over campus. It was a great fit for my research (engineering communication), and the small size seemed like it would lend itself very well to collaborative opportunities with faculty from other disciplines. I didn't give too much thought initially to the status of the Technical Communication program as the only liberal arts-based degree within the university. I quickly learned however, that unlike Technical Communication faculty housed in traditional English departments, who sometimes have the challenge of defending our discipline as legitimate and based in rhetorical tradition rather than vocational, I faced the challenge of presenting our discipline as applied and relevant to the engineering and scientific work happening all around me, yet still grounded in an age-old rhetorical tradition.

It took just a few months' worth of faculty senate meetings, participation in a campuswide assessment committee, and a workshop I delivered to engineering faculty concerning how to integrate communication instruction within engineering courses (motivated by an ABET reaccreditation visit happening the following year) for me to discover that the risk of a liberal arts–based department at a technical institution is that you (or your courses) may be viewed suspiciously by engineers and scientists as being "touchy-feely," a fate worse than death in a technical and scientific environment. Fortunately, the value of communication instruction for students is well understood by most science and engineering faculty on our campus. I did figure out that it is important, however, that this instruction not be packaged in a manner fraught with touchy-feeliness.

## What is an Assistant Professor of Technical Communication to Do?

Serving as Technical Communication Program Director while on the tenure track in a science and engineering climate in which research expectations are high required a lot of prioritizing, boundary-setting, multitasking, and efficiency. Coffee helped immensely too. Although I was initially granted a course release (reducing my teaching load to a 3/2), after the first 2 years on the job, my teaching load returned to 3/3. Realizing that in order to perform my duties as program director and achieve tenure I would need help, I involved my departmental colleagues as much as possible in the program. My efforts here were also motivated by the belief that the more exposure and contact students could have with individuals from neighboring disciplines, the fuller and richer their academic experience would be. In addition to having them serve as advisors to Technical

Communication students, I invited colleagues to consider cross-listed courses that could satisfy our program's requirements and also expose students to approaches toward communication from outside fields. I mentored willing faculty members in Composition who taught Technical Communication courses for the first time. I also took advantage of the corporate advisory board's willingness to be more involved in our program through initiating activities such as pairing senior thesis students with corporate board mentors (for more information about our corporate advisory board's role in the program, see Ford, Bracken, & Wilson, 2009 and Ford et al., 2013 forthcoming).

Lest I create the impression of possessing superhuman abilities or being a workaholic, let me clarify that I do not share the tasks that program directors in larger settings do. I do not have to oversee TAs (we do not have a graduate program). I do not have to hire or supervise adjuncts (the few adjuncts our department has teach first-year composition and are overseen by the Writing Program Coordinator), and I am not responsible for controlling a budget (the only positive aspect about not actually having a budget). As Catherine Latterell explained in her 2003 *Journal of Technical Writing and Communication* article about technical and professional communication programs within small college settings, "In small schools, one learns that there are very few opportunities for wielding real power. . . . As such, establishing one's good will and operating through personal influence can be more meaningful than the having or keeping of titles" (p. 332).

As I was working on establishing my own research agenda in engineering communication, I began to cultivate some relationships outside the department. These relationships not only helped move forward my path toward tenure, but they were key to establishing the good will Latterell (2003) referred to and in helping to secure the Technical Communication program's status as valued within the institution. Reaching out to colleagues in engineering and scientific disciplines was not terribly difficult. As Latterell noted, "Collaborations that could be difficult for larger programs to undertake can be more possible at small schools due to the interdisciplinarity and flexibility that pervades" (p. 329). Further, "the nature of decision-making at small schools is inevitability compressed and always involves cross-disciplinary negotiation" (p. 331). For example, I didn't feel too uncomfortable "cold calling" the professor teaching a junior design course in the Electrical Engineering department to discuss a possible research collaboration.

Although Latterell's (2003) argument concerned recognizing "opportunities to integrate technical and professional communication studies within the arts and humanities," my challenge was to integrate technical and professional communication with engineering and science (p. 328). Fortunately, I understood the climate I was working in well enough to know that as long as the initiatives I was proposing were ones that reflected a Writing in the Disciplines approach, most faculty would be receptive.

It didn't take long for invitations to work with other disciplines on campus to come my way. As Technical Communication Program Director, I had many requests from science and engineering colleagues to contribute to discussions and activities from other disciplines involving communication instruction. In turn, I could very much relate to Latterell's (2003) noted worry "about being pulled into literally every programmatic initiative across the campus" (p. 328). Although these invitations to serve on committees, participate in assessment activities, and so on, meant additional service obligations, I also viewed them as chances to elevate the status of our program across campus as well as attain resources for students. With my background and research interests in engineering communication, I began to give priority to the requests coming from engineering departments, in particular the largest engineering department on our campus, Mechanical Engineering. My interaction with the Mechanical Engineering department has taken several different forms. In the next section, I describe the activities I undertook and the ways in which our Technical Communication program has benefited.

## MECHANICAL ENGINEERING AND TECHNICAL COMMUNICATION

For almost a decade, the enrollment in engineering majors at New Mexico Tech has been steadily increasing. Mechanical Engineering (MENG) in particular has exploded, now boasting the distinction of our largest undergraduate major on campus, with 313 students. The number of MENG students awarded degrees more than quadrupled from 2004 to 2012.

In 2005, the Mechanical Engineering department began making a dedicated effort to go beyond the general degree requirements of the required first-year composition sequence and a junior-level technical writing course to improve MENG student communication skills. The MENG faculty noticed deficiencies in student writing and speaking skills, and they were also responding to information from alumni, employers, and the Accreditation Board for Engineering and Technology (ABET) attesting to the necessity for clear and effective communication skills. The department chair and I had discussions about possible ways the Technical Communication program could help. Initial activities included me offering technical writing course refreshers to students in the design clinics, attempting to fill a section of the general education technical writing course with mostly MENG students, and holding additional office hours specifically for MENG students.

I carried out these activities over the course of 4 years, and they were successful in strengthening student communication skills (anecdotal evidence from the faculty teaching the junior and senior design clinics pointed toward better quality presentations and reports, and all department faculty noted improved thesis writing and presentations among students who had taken the graduate-level course). The MENG department was pleased with the outcome and wanted to increase my participation. In 2009, another opportunity came. Our university was awarded a Title V Promoting Postbaccalaureate Opportunities for Hispanic Americans (PPOHA), grant funded by the Department of Education. Increased communication instruction was a main emphasis of that grant, and as a result, I was presented an opportunity to develop and teach a graduate-level course in communication for Mechanical Engineering.

After that point, I was more than just a casual participant in the MENG department. In 2011, I was invited to move into a joint appointment position in which I would split my teaching and service duties 50/50 between the Communication, Liberal Arts, Social Sciences department and the Mechanical Engineering department (please see my forthcoming *Composition Forum* article for more details about the evolution of this joint appointment).

Through the joint appointment, I have not only been able to impact MENG curriculum from *within* the department, which I believe is extremely beneficial for communication instruction, but I am also in the unique position of being able to promote and establish ties between the Technical Communication program and the largest and fastest-growing undergraduate major on our campus. Further, my involvement in teaching graduate-level communication courses to MENG students is the first step toward a graduate certificate program, something the Communication, Liberal Arts, and Social Sciences department colleagues are interested in establishing. Following, I provide more information about each of these initiatives.

## Interdisciplinary Initiatives

As Stuart Blythe noted in his 2000 CPTSC conference proceedings, technical communication students should possess competencies beyond writing, and there is much value in seeking interdisciplinary models. Although Blythe pointed out that "these competencies fall outside the disciplinary boundaries as defined at our local institutions and in fact we may face penalties for developing such competencies" (p. 35), fortunately the environment at New Mexico Tech supports rather than penalizes interdisciplinary efforts.

It also does not hurt that interdisciplinary initiatives help enact and support ABET objectives and the suggestions offered in the *Educating the Engineer of 2020* report, published by the National Academy of Engineering (2004), which seeks to transform learning environments to guarantee the success of engineering students. The authors of this report recognize that there is not one right way to achieve this goal, but instead, "campus leaders must exercise leadership to shape an agenda for action that makes sense for them, given their mission, circumstances and vision of the future" (p. 20). Given my allegiance to the Technical Communication program, transforming our campus learning environment in a manner that joins Technical Communication with Mechanical Engineering makes the most sense. Although no "Technical Communicator of 2020" report exists, our field is well aware of the value of interdisciplinary learning. As Paretti (2011) explained, an interdisciplinary framework combining both areas of expertise is ideal for rhetorical instruction. The concept of interdisciplinarity is summarized by Paretti as one that "operates around shared goals that reside at the intersection of disciplinary boundaries," wherein "collaborations are marked by epistemological openness and a recognition of the value of each perspective" ("interdisciplinarity").

The Technical Communication students have been connected with Mechanical Engineering design clinic students through activities such as help with proposal writing (the Persuasive Communication students collaborated with a design clinic team by helping them plan, draft, and prepare a funding proposal to Google). Another example of this kind of collaboration involved the Persuasive Communication students working with a design clinic team on their marketing materials (that design team was seeking additional sponsorship and needed brochures that explained their project in terms nonengineers could understand.)

As I continue to be involved with the MENG design clinic, the prospects to link Technical Communication class assignments and activities to design clinic projects will be many (with 17 different teams working on design projects for external clients, opportunities abound), and I plan to take advantage of them. For example, one of our required Technical Communication courses is a senior-level course titled Professional Writing Workshop. The title and description of this course is intentionally vague so that we can shape the course toward a particular project or collaborative opportunity. When we offer this course again in spring 2013, the plan is to tie it to the Mechanical Engineering design clinic. Like Aller and Clancey (2000), who in their CPTSC 2000 conference proceedings presented a model at Michigan Tech in which they had the opportunity (and challenge) of developing and integrating communication initiatives within modules in a design clinic, in future efforts, I will continue to plan initiatives with the goal of "maintaining pedagogical and programmatic integrity" while responding to initiatives to change curricula to better satisfy National Academy of Engineering and ABET objectives (p. 80).

Another idea for the future is a cross-listed Technical Communication/ Mechanical Engineering course that involves ergonomics, an area in which MENG advisory board members and alumni have specifically requested instruction. The close tie between ergonomics and usability makes for a fantastic opportunity to offer a course suitable for both disciplines.

## Future Grant Opportunity

Currently, I am working on a future grant initiative to begin an engineering leadership program that will include communication as a central component. Although the idea is in its infancy right now, I do envision a Technical Communication course focused on leadership and collaboration as one of the centerpieces of the program. This initiative is driven by concerns regarding students' preparedness in leadership, collaboration, and networking voiced by faculty and advisory board members at our recent Mechanical Engineering program advisory board meeting. Coincidentally, at our last advisory board meeting for the Technical Communication program, our board members also raised questions regarding our Technical Communication students' preparedness in the areas of networking and serving in leadership roles. An ideal program then would pool resources and be shaped by both departments and serve both Technical Communication and Mechanical Engineering students.

## Opportunities via Advising

Part of my service duties within the Mechanical Engineering department include serving as an academic advisor to MENG students. I currently have over 30 MENG advisees. In working with these students to help them choose courses each semester, I seize opportunities to steer them toward Technical Communication courses to fulfill their required Humanities credits. Especially when students have no idea what Humanities classes they want to take, I point out the close connection between courses like our Visual Communication and Persuasive Communication courses and the visual display and proposal-writing tasks they will be required to complete in their future engineering courses and in the workplace. The result is that I help to ensure that these technical communication courses maintain full enrollments (which preserves our good standing in the eyes of administration), and MENG students get valuable exposure to Technical Communication courses. Several of them end up taking a second Technical Communication course, and we have had a handful of MENG students go on to declare a minor in Technical Communication. Further, our Technical Communication majors in those classes benefit from working in group activities with students outside their discipline.

## Internship and Community Service Opportunities

In addition, all Technical Communication students must complete an internship requirement and a community service requirement. Both requirements can be fulfilled by partnering with clients on campus; so for students wanting on-campus experiences, there are options for linking them with projects stemming from the Mechanical Engineering department. Such an experience would provide students with great exposure toward engineering projects, providing valuable "real-world" preparation.

## Graduate Certificate Program

About a year ago my colleagues in the Communication, Liberal Arts, and Social Sciences department started discussing the possibility of adding a

Graduate Certificate Program in Technical Communication. Although we have kicked around the idea of proposing a master's of science program for many years, our lack of faculty and high teaching loads and service obligations has always prevented us from getting beyond the discussion stage. A graduate certificate program, however, would be a much more feasible goal and could certainly begin paving a path toward any future hopes of a master's program. Because, as a department, we are already offering a few graduate seats in some of our Technical Communication, Communication, English, Philosophy, and Anthropology courses (thanks to the Department of Education Title V PPOHA grant mentioned earlier), putting a graduate certificate program in place would not be too difficult to achieve.

Promoting this Graduate Certificate program among the Mechanical Engineering graduate students would be one of my goals. A few models for graduate certificate programs in technical communication aimed toward engineering students already exist, such as Purdue and the University of Wisconsin (Reaves 477). Because our university requires all graduate students to take at least two courses outside their major discipline, there is already incentive for MENG students to take at least one Technical Communication course. The graduate-level Communication in Mechanical Engineering course that I developed initially as part of the Department of Education Title V PPOHA grant activities is now offered each spring and is being promoted heavily among MENG faculty. A graduate certificate program would ensure that there are regularly taught Technical Communication courses with seats set aside for graduate students, so it wouldn't be too hard to imagine an MENG student taking both their required two courses outside the department in Technical Communication. For those students who showed interest in taking more, adding the two or three more courses it would take to constitute a graduate certificate is not out of the question.

In addition to serving as a first step toward us developing a master's of science program in the future, the Graduate Certificate Program would also provide other benefits. It would help ensure healthy enrollments in our classes, which translates into good standing in the eyes of administration and could further our cause as we argue for new faculty lines, increased programmatic resources, and so on. Further, the revenue from teaching graduate-level students would increase our department's FTEs, which again, helps better our chances for adding tenure-track faculty and securing more resources. Most importantly though, the addition of graduate students from engineering or scientific disciplines in our Technical Communication classrooms would provide valuable exposure for our undergraduate Technical Communication students. Hearing perspectives from advanced students from technical disciplines and collaborating with them on class projects and activities would grant them experiences that would help prepare them for the technical and scientific workplaces they will later enter.

## CONCLUSION

In her 2010 *IEEE Transactions on Professional Communication* article, Marjorie Davis noted the value of technical communication faculty working within engineering contexts, claiming that "the rewards for this intentional engagement benefit not only the engineering and technical communication faculty but also their students and the students' future employers" (p. 44).

In her portrait of Mercer University's integrated technical communication and engineering programs, Davis advocates for technical communication faculty to serve as "an integral part of the educational experience instead of remaining detached from the educational goals and practices within engineering" (2010, p. 44). Although Mercer's Technical Communication program is officially housed within their School of Engineering, the strategy Davis advocates for guaranteeing successful teaching of technical communication to engineering students is one that we have been able to adapt at New Mexico Tech. This strategy is to provide "multiple opportunities to communicate, especially within real-world work contexts," rather than limit the teaching of technical communication to a "one-shot experience not well integrated into the entire engineering education program" (p. 34).

As the initiatives I have detailed in this chapter suggest, I am in firm agreement with integrating technical communication and engineering disciplines as extensively and as fully as possible. At New Mexico Tech, I have the freedom and support to be able to make these intersections happen, and for that I am grateful. Although I have offered the benefits of each of these initiatives throughout this piece, I would like to close by highlighting one final overarching benefit. In the National Academy of Engineering's (2004) *The Engineer of 2020 Report,* creativity is treated as "an indispensable quality for engineering" (p. 55). The ability to invent, innovate, and think outside the box is necessary for future engineers who will face a "growing scope of the challenges ahead and the complexity and diversity of the technologies of the 21st century" (p. 55). This emphasis on creativity is not essential just for engineers, but technical communication students will face similar challenges and experience the same complex and diverse 21st century technologies. To help foster this creativity, both technical communication and engineering students should be afforded the opportunity to approach problems from an interdisciplinary perspective and collaborate with others outside their discipline.

## REFERENCES

Aller, B., & Clancey, M. S. (2000). Creating communication modules for an engineering enterprise initiative: Programmatic and rhetorical considerations. *CPTSC Conference Proceedings,* 80–81.

Blythe, S. (2000). The value of seeking interdisciplinary models for smaller professional writing programs. *CPTSC Conference Proceedings,* 69–70.

Davis, M. T. (2010). Assessing technical communication with engineering contexts. *IEEE Transactions on Professional Communication, 53*(1), 33–45.

Ford, J. D. (2012, forthcoming). Integrating communication into engineering curricula: An interdisciplinary approach to facilitating transfer at New Mexico Institute of Mining and Technology. *Composition Forum.* Retrieved from http://compositionforum.com/issue/26/new-mexico-tech.php

Ford, J. D., Bracken, J. L., & Wilson, G. D. (2009). The two-semester thesis model: Emphasizing research in undergraduate technical communication curricula. *Journal of Technical Writing and Communication, 39,* 433–454.

Ford, J. D., & Lanier, C. R. (2011). If you build it they will come: Establishing a research group at New Mexico Tech to increase campus visibility of the technical communication program. *Programmatic Perspectives, 3*(1), 96–106.

Ford, J. D., & Newmark, J. (2011). Emphasizing research (further) in undergraduate technical communication curricula: Involving undergraduate students with an academic journal's publication and management. *Journal of Technical Writing and Communication, 41,* 311–324.

Ford, J. D., Newmark, J., & Durao, R. (2013, forthcoming). Writers among engineers and scientists: New Mexico Tech's bachelor of science in technical communication. In G. Giberson, J. Nugent, & L. Ostergaard (Eds.), *Undergraduate writing majors: Nineteen program profiles*. Logan, UT: Utah State University Press.

Latterell, C. G. (2003). Technical and professional communication programs and the small college setting: Opportunities and challenges. *Journal of Technical Writing and Communication, 33*(4), 319–335.

National Academy of Engineering. (2004). *The engineer of 2020: Visions of engineering in the new century.* Washington, DC: National Academies Press.

Newmark, J., & Ford, J. D. (2012). An academic ejournal as technical communication client project: Enculturation, production, and assessment. *Technical Communication, 59,* 286–301.

Paretti, M. (2011). Interdisciplinarity as a lens for theorizing language/content partnerships. *Across the Disciplines: A Journal of Language, Learning, and Academic Writing, 8*(3). Retrieved January 5, 2012 from http://wac.colostate.edu/atd/clil/index.cfm

CHAPTER 11

# Expertise in Professional Communication as a Catalyst of WAC/WID Administration Success

*Pavel Zemliansky*

## INTRODUCTION

In recent years, professional communication scholars have noticed substantial similarities and connections between their field and writing across the curriculum and writing in the disciplines (WAC/WID). For instance, in 2007, *The Journal of Business and Technical Communication* devoted an entire issue to writing in the disciplines. The articles included in that issue argue that the theories and practices that guide professional communication can be successfully applied to the teaching of writing in the disciplines and, by extension, to the administration of WAC/WID programs. Among such principles are the contextuality and disciplinarity of writing, the importance of both "writing to learn" and of "learning to write," and others. I use two of the articles from that special issue to frame my narrative.

This chapter presents a narrative of my work as the Director of Writing Across the Curriculum at the University of Central Florida. I describe the key activities of the program during my first year as its director and reflect on how, as a WAC director, I have followed and applied (sometimes consciously and sometimes not) principles and approaches that I had learned and practiced for many years as a teacher of professional communication and as an administrator of a professional communication program.

The segments of the narrative are framed in some key ideas from the fields of professional communication and WAC/WID. These frames are not literature reviews but rather an attempt to highlight the key concepts relevant to the use of professional communication theory and practice in WAC/WID work. Such theoretical grounding of my narrative allows me to heed the call of the editors of this collection to "recast administration as a rich, complex intellectual engagement rather than an activity based only in institutional details and processes" (Bridgeford, Kitalong, & Williamson, this volume, p. 3).

## THE WRITING ACROSS THE CURRICULUM PROGRAM AT UCF: INSTITUTIONAL CONTEXT

Program administration is a highly situated activity whose success and day-to-day operations depend on the local conditions on the ground. Because of the importance of local contexts for successful administration, information about the institutional context of UCF and of its WAC program is in order.

To date, UCF is the second-largest institution of higher learning in the United States, with 58,698 students, 1,948 faculty members, and 11 campuses across central Florida (Institutional Knowledge Management, 2012). In 2012, UCF offered 92 undergraduate and 118 graduate and professional programs. The Carnegie Foundation for the Advancement of Teaching classifies UCF as a research university with very high research activity (Institutional Knowledge Management, 2012). UCF's student body is a combination of traditional students who live on or close to campus and go to school full time and nontraditional and commuter ones who combine college with full- or part-time work and family life. Over 95% of UCF's students are from Florida, which reflects the university's commitment to focusing on providing high-quality educational opportunities to residents of Central Florida and the rest of the state.

Every year, UCF accepts many transfer students from both other 4-year institutions and from Florida's 2-year colleges. In the summer of 2011, when I was hired, the university took in 6,000 transfer students (Institutional Knowledge Management, 2012). The data on transfer students is important because having so many transfer students on campus, while good for student body diversity, ensures a lack of uniformity in those students' previous experiences with academic and professional writing instruction. This lack of uniformity creates a set of interesting challenges for the writing across the curriculum program.

The WAC program at UCF is housed in the Department of Writing and Rhetoric. The program's staff consists of a WAC Director, responsible for the overall development and running of the program; a WAC Coordinator, who assists the director with a variety of projects; and a part-time administrative assistant. Both the WAC Director and WAC Coordinator are also regular faculty members in the department, at the rank of associate professor and full-time instructor, respectively.

### Using Professional Communication Expertise to Develop WAC Programs

Concerted and deliberate efforts to deliver writing instruction as part of disciplinary curricula can, among other things, help improve "student's domain-based learning through writing instruction" (Russell, 2007, p. 253) and to "socialize students into their discipline" (Russell, 2007, p. 258). In addition, several professional communication scholars have demonstrated how methods and approaches from professional writing can translate to improved efforts in WAC/WID. Recognizing that much of the writing instruction in the disciplines is about socialization of students into those disciplines, Carter, Ferzli, and Wiebe (2007, p. 268) demonstrate that students who wrote as part of a biology lab course improved their level of socialization into their academic discipline (p. 268).

Professional communication as a field has created a rich understanding of how writing is learned and practiced, both in academia and in the workplace. Russell (2007) states that "in recent years, [business and technical communication] teachers and researchers have looked closely at the discourse and activity of the disciplines, and the professions beyond them, to formulate curriculum and pedagogy" (p. 249). Because of this expertise, professional communication specialists are often the people to whom other members of a university community turn for help with student writing and who end up being charged with the development of various WAC/WID initiatives.

Russell's (2007) review of research on the connections between business and technical communication (BTC) with WAC/WID demonstrates that many WAC/WID initiatives turn to professional writing courses, hoping to fill the gap in the writing preparation of students in other disciplines (p. 250). Because courses in business and technical communication are "more focused" than general education first-year writing courses, they are more suitable for WID work (p. 250). The collaborations between professional writing and other faculty, aimed at improving writing instruction, can take many forms, ranging from joint assignment course and program design and delivery to workshops, seminars, and other consultations by writing faculty for colleagues in other disciplines.

Much of the research I summarize above deals with the models of WAC/WID, which use the expertise of business and technical communication faculty to design and deliver writing courses in other disciplines. Although such work can yield good results in some settings, I am more interested in the work of WAC/WID specialists as consultants to other faculty. WAC programs that adopt the consultant model do not directly design and deliver writing courses for other departments and programs but advise colleagues on the creation of approaches to writing instruction, which fit into their disciplines. The WAC program at UCF has chosen the consultant model as well.

As the modern workplace changes, so does the role of a professional communicator. That role has shifted from a content producer to project manager and consultant. Professional communicators' work has changed from solitary to social. Modern professional communicators "manage situations, events, and the interactions and experiences of individuals, groups and cultures" (Moore & Kreth, 2005, p. 303). In these new roles, professional communicators are essentially consultants who "add value to their organizations" (Moore & Kreth, 2005, p. 303).

Conklin's study of the extent to which modern technical communicators are focused on the creation of texts versus "facilitating collaboration and participative processes" found that "technical communicators are constructing a shared practice that is perceived as contributing significant value to a successful technology based organization" (2007, pp. 211, 228).

Artemeva (1998) recounted her experience of consulting for two engineering firms, one from Canada and the other from Russia. The two first were engaged in a joint engineering project, and the consultant's role was to ensure successful communication between participating engineers from the two countries. Specifically, Artemeva's work as part of this project included consulting both sides on cultural, linguistic, and other relevant issues and facilitating the production, by the Russian engineers, of periodic engineering reports for their Canadian clients (p. 285).

In their role as consultants, professional communicators often have to help various individuals and organizations to work together and to create new meaning by reconciling professional, disciplinary, and other differences. We at UCF also work with faculty in other disciplines, helping them to add value to their programs by improving the learning and teaching of writing.

For any consulting project to be successful, relationships between the consultants and the clients need to be established and maintained. Such relationships need to depend on and utilize the expertise and interest of both the consultant and client. Having such mutually beneficial relationships will not only replicate the consultant-client dynamic of the professional world, but also help programs avoid slipping into the "conversion model" of WAC, "which assumes that faculty in other disciplines are heathens who must be converted to the Right Way" (Walvoord, 2000, p. 11).

## WAC AT UCF: INITIAL STEPS

Before my arrival at UCF during the 2010–2011 academic year, my colleague Elizabeth Wardle had started some excellent WAC pilot projects with the Department of History and the College of Nursing. Both projects consisted of training and "embedding" tutors into courses in those two units. The tutors worked with the course instructors to help students become better writers in the disciplines. These projects not only provided the much needed "first push" for the WAC program but also gave me a small but dedicated cohort of "early adopters" of WAC from those two departments with whom I could begin expanding WAC efforts right away.

During the fall semester of the 2011–2012 academic year, the activities of the WAC program proceeded in two directions. We worked to maintain existing relationships with the History and Nursing faculty, who had participated in the pilot projects during the previous year, while developing new relationships, connections, and areas of collaboration throughout the university.

Successful department-centered WAC work depends on the understanding by WAC faculty of the specific characteristics of writing and of the teaching of writing in other disciplines. To be effective consultants to faculty in other departments, we need to go beyond the comfort zone of our own fields, whether it is professional writing, rhetoric, or composition, and learn about how writing is practiced, learned, and taught in other disciplines. Otherwise we are in danger of replicating what Russell (2007) called a way of disciplinary writing instruction, uncritically modeled upon first-year composition courses. According to Russell, such a model engenders a "conceptual separation" of writing instruction and the disciplines in which it takes place, thus drastically reducing its effectiveness (p. 250).

Understanding this, I immediately visited with the History and Nursing faculty who had participated in the pilot projects during the previous year. The purpose of these visits was not only to meet my new colleagues and offer them ongoing WAC support but also to learn about their departments' curricula, their perceptions of the quality of their students' writing, and of their needs in relation to WAC. Many of the projects implemented during the 2011–2012 academic year came out of those meetings.

The next few pages of the chapter describe the early work of our program, designed to strengthen and deepen existing relationships while preparing ground for the building of new connections with faculty. This description is chronological and focuses on early WAC workshops and other events we organized in an effort to attract more attention to WAC work at UCF. Then comes the section entitled "The Department-Centered Model of WAC and the WAC Fellows Program at UCF." That section begins with a theoretical foregrounding of the WAC Fellows initiatives before the chapter resumes its chronological structure.

Continuing to work with our early adopters was fine, but we also needed to expand our circle of contacts on campus. To achieve that goal, we decided to conduct several university-wide workshops on WAC topics, inviting all and any interested faculty to them. We weren't breaking any new ground here because the WAC workshop is a time-honored way to get faculty from across the disciplines to discuss the teaching of writing. We certainly understood that we would not be able to change anyone's minds about writing instruction in the disciplines with one or two workshops. Therefore, we saw these workshops as an opportunity to meet new people and to engage those people in conversations about writing across the curriculum.

During conversations with colleagues earlier on, the problem of large class sizes came up repeatedly. We quickly realized that because so many of UCF's faculty teach large classes, they might perceive WAC as a "mission impossible." Therefore, it was almost natural for us to devote our first workshops to the teaching of writing in large classes.

Earlier on, we had decided to follow the "learning first, writing second" model of WAC, in which writing is taught and learned as a means to learn something about the student's discipline or discourse community. This "something" could be a piece of disciplinary content, a way of thinking about that content in a new way, or learning how to write in disciplinary genres. Because of that focus on learning as the ultimate goal of writing instruction, we decided to call our first workshop "Increasing Student Learning in Large Classes Through Writing."

Our rationale for coming up with this title was as follows: some faculty members may be quite skeptical about the very possibility of teaching writing in their courses or may think that, because of the size of their classes, writing instruction is not for them. However, everyone (or almost everyone) who teaches cares about their student learning. Thus, we were attempting to offer writing not as a self-contained activity but as a means of achieving a greater level of student learning in large classes. Anticipating skeptical reactions of teachers of large classes to the proposition of including lengthy, formal writing projects in their courses, we decided to focus the workshop on introducing faculty to the notion of writing-to-learn and to informal, low-stakes writing assignments that could be assigned in large classes and assessed relatively painlessly.

To accommodate the largest possible number of participants, we scheduled two sessions of the workshop on 2 consecutive days: one in the morning and the other in the afternoon. With a PowerPoint presentation and multiple copies of a lengthy workshop handout, we were ready to help multitudes of faculty members from different departments with the virtues of writing-to-learn assignments and the benefits that assigning writing-to-learn has for their students' learning. However, when we arrived at the workshop, we discovered that we had only one attendee, a member of the History Department. Naturally we had to radically alter our plans and turn the workshop into a more informal question-and-answer session and conversation with that teacher. The next day, the attendance situation was similar: one faculty member attended the workshop, this time from the Department of Chemistry. Like the previous day, our well-structured and rehearsed workshop scenario had to be abandoned in favor of a more unstructured conversation about teaching writing and critical thinking.

We were disappointed by such low attendance. Looking back, the low attendance at those first two workshops may have been a blessing in disguise because it gave us time to devote undivided attention to the conversations with the teachers who did attend and to understand their interests and concerns in more detail. The opportunity to spend some one-on-one time with those teachers helped us to position ourselves as consultants who would work with specific clients on specific teaching problems in their specific courses and disciplines. In that setting, the danger of the "one size fits all" and "sage on the stage" effect, which can creep into even the best-intentioned and designed workshop, went away and was replaced with the spirit of genuine conversation among colleagues concerned with student learning.

Later, the history and chemistry teachers who attended our first workshops became instrumental in convincing other colleagues in their departments to join our flagship initiatives, the WAC Fellows Program, which took place during the spring semester of 2011–2012. Moreover, due to her senior position in the department, the chemistry professor has been working with her department chair and other colleagues to introduce some substantial changes to several chemistry courses following her work with us. The WAC Fellows Program is our largest and most successful effort to date, and it is detailed in the next section of the chapter.

The first round of workshops had made it very clear to us that, due to the large size of many classes at UCF and due to a variety of other pedagogical concerns, the topics of writing-to-learn and of low-stakes writing assignments to improve student critical thinking and learning would be crucial for our efforts to attract and sustain faculty interest in WAC. While we certainly did not and still do not discount the importance of learning-to-write assignments, we are finding that one path toward convincing faculty in other disciplines to assign more writing of different kinds in their courses is by helping them understand how regular practice in writing can help students improve their learning and thinking as members of their academic discipline. It was also important for us to show that, although faculty in other disciplines are not writing specialists, with the right amount and kind of training provided by the WAC program, they could design and teach high-quality writing assignments that would result in improved student learning.

Therefore, during the remainder of the fall 2011–2012 semester, we conducted workshops on two more topics: "How Low-Stakes Writing Assignments Can Help Students Learn and Save Teachers Time" and "Responding, Assessing, Grading: How to Provide Useful Feedback to Student Writing." The attendance at those workshops improved somewhat, with 3–4 faculty members attending each of the sessions. And although such attendance rates could hardly be called successful for a university-wide workshop, especially at an institution of UCF's size, as I stated earlier, having the ear of a few dedicated and interested colleagues who did show up at the workshops allowed the WAC program to begin building in-depth consultant-client relationships with those faculty members and their departments, if not achieving the breadth of WAC coverage by means of the workshops. For breadth of coverage and for increasing our footprint on campus, we had other plans and other initiatives in mind. The first and most important of these initiatives, the WAC Fellows Program, is described in the next section of the chapter.

## THE DEPARTMENT-CENTERED MODEL OF WAC AND THE WAC FELLOWS PROGRAM AT UCF

As I mentioned earlier, before proceeding with a chronological account of the development of the WAC program at UCF, it is necessary for me to foreground my narrative in an important theoretical concept: the department-centered model of WAC. In his keynote address to the 11th International Writing Across the Curriculum Conference, Chris Anson (2012) suggested the department-centered model of WAC may be both the most sustainable and the most effective way to develop WAC projects. In the department-centered model, the responsibility to shape and implement WAC activities is placed on individual departments and units. With such responsibility comes the opportunity to shape the methods and approaches of writing instruction in the department, including the design and methods of delivery of specific courses and assignments. According to Anson, who directs the highly successful Campus Writing and Speaking Program at North Carolina State and whose program is also based on the department-centered model, the role of WAC specialists is to consult and advise their colleagues in other departments on the best ways to teach writing in their own disciplines. In addition to his own successful WAC program, Anson mentioned the programs at UNC Charlotte, the University of Minnesota, and other institutions that utilize the same model. His talk included testimonials by faculty from those institutions who told of their successes with WAC.

The WAC program at UCF follows the department-centered model. Our WAC projects are close partnerships between faculty members in other disciplines and WAC faculty. During these partnerships, WAC faculty provide colleagues in other disciplines with both general knowledge about writing studies as a field and help in connecting that knowledge with the specific needs and circumstances of their fields, departments, and students. We believe (and research supports our belief) that such a model creates a vibrant and sustainable WAC effort.

In order to implement the department-centric model, we have focused our work with departments on the development of writing-related student learning outcomes and of writing assignments in support of those outcomes. These outcome sets are both framed around the disciplinary conventions of a given department or program and informed by the latest theory and practice from WAC. The process works in the following way: faculty from participating departments provide us with the "raw material" necessary for the articulation of the outcomes, and the WAC program consults with them on such articulation, helping them infuse their thinking about writing in their discipline with WAC concepts, research, and terminology. The end results are always designed to be practical and usable in the classroom.

Once faculty in other disciplines have drafted their department-specific learning outcomes for writing, the WAC program helps those departments design formal and informal writing assignments in support of those outcomes. During the 2011–2012 year, which was the first official year of the WAC program's existence, we collaborated with seven departments and programs on the creation of those outcomes and related writing assignments.

To achieve these goals, in the fall of 2011 we decided to create a program that would allow faculty to work on the creation of department-specific writing learning outcomes and assignments in support of achieving those outcomes. We chose to construct a sustained and structured program that would bring together several small teams of dedicated faculty from different departments. Those teams would work with us and with one another on developing writing outcomes and assignments for the programs. We wanted this program to last at least a semester, providing participating faculty members opportunities both to work with the WAC faculty and also to learn from one another. Each participant would receive a $1,000 stipend upon completion of the semester-long program and submission of all required deliverables. As a result, the WAC Fellows Program at UCF was conceived in the fall of 2011.

The website of the WAC program at UCF describes the WAC Fellows project in the following way:

> The WAC Fellows Program is a combination of large and small group meetings. During the plenary meetings, the participants discuss various topics of writing theory and pedagogy and share with each other their writing instruction experiences, plans, and the materials they are developing. During small-group meetings, departmental teams work on the creation of those materials. (UCF WAC Program, 2011)

The WAC Fellows participate in large and small group meetings, during which they collaborate with each other and with the WAC faculty on the creation of the required deliverables.

The first session of the WAC Fellows Program took place during the spring semester of 2012. Participants included 11 faculty members from History, Nursing, and Chemistry. During the first running of the program in the spring of 2011, the WAC Coordinator and I held four plenary meetings, which were attended by all 11 participants. At the plenary meetings, we gave brief presentations about key topics of writing theory and pedagogy (disciplinary conventions of writing, responding to student writing, writing-to-learn, and others) and discussed the core text of the program, John Bean's excellent book, *Engaging Ideas* (2011). Participants also had the opportunity to share ideas and stories about the teaching and learning of writing in their disciplines and departments. During the last two plenary meetings, after the work on learning outcomes and assignments began in earnest, an arena for the three departmental teams to share their work in progress with each other was provided.

Between the plenary meetings, we met with each departmental team individually. During those meetings, we answered questions and addressed concerns about their department-specific projects. We also helped participants articulate specific tasks and deliverables that contributed to the overall mission of the project. Departmental teams then completed those tasks on their own, with the results of that work being presented and discussed during subsequent small-group and plenary meetings.

We repeated the WAC Fellows program in May of 2012 (summer session at UCF, officially). To accommodate summer schedules of faculty, we decided to change the program's format to an intensive, four-day event, during which 18 faculty members from the Departments of Psychology, Mathematics, Electrical Engineering and Computer Science, the School of Communication, and the College of Hospitality Management worked to create outcomes and assignments for their departments. Each day of the program included large and small team meetings, presentations, and conversations, so overall, the program remained true to its original goals and procedures.

The WAC Fellows Program has so far contributed to the development of the WAC program at UCF in some important ways. First, it helped us advance our vision of the consultant and department-centered models. With our help, the WAC Fellows developed discipline and department-specific sets of writing outcomes and assignments. Second, and on a broader level, the WAC Fellows Program has contributed to the popularization of the idea of WAC at our university and to the intangible project of building relationships, networks, and "circles of influence" between the WAC program on one side and faculty and administrators in other departments on the other. So far, this popularization of WAC has happened in a couple of ways. The showcase event, which took place at the end of the spring Fellows program was attended by department heads and associate deans from the participating units. Also, faculty members who did not participate in the Fellows programs but who came to us for writing consultations have been able to learn about outcome sets and writing assignments developed by the WAC Fellows and to adjust those for their own needs and their own courses. One example of such use is the "repurposing" of the Undergraduate Research Report assignment developed by the WAC Fellows from Chemistry, by a professor from UCF's NanoScience Technology Center for the needs of an NSF-funded undergraduate research initiative, which he runs.

The Fellows program has also been providing us with lessons about what works and what doesn't, even with a glimpse of what we might expect in the future. We began the first plenary meeting of both the spring and summer WAC Fellows Program by asking participants to create a list of characteristics of an "ideal student writer" in their disciplines. The lists that the representatives of these different disciplines (ranging from Psychology to Mathematics and Computer Science, to Hospitality Management) came up with were remarkably similar. Virtually all participants valued such qualities as critical thinking, ability to conduct research responsibly and ethically, ability to address professional and academic audiences appropriately, and so on. So despite the importance of the differences in academic and professional writing conventions and processes between different academic disciplines, professions, and departments, we learned from listening to and participating in these exchanges that our client departments at UCF are concerned with a lot of the same issues. Equipped with this knowledge, we began to understand how to structure future development of the WAC program at UCF. In this work, we need to, on the one hand, keep those disciplinary differences in mind and to avoid the "one size fits all" approach and, on the other hand, not to lose sight of the overriding reason for the existence of the WAC program, and that is to improve student learning through better writing instruction.

## ESTABLISHING A WEB PRESENCE: THE WAC@UCF WEBSITE

In his article, "WAC Websites as Knowledge Webs," Dan Melzer (2006) argued that often, creating a website for a WAC program is an "afterthought," something that is done after the WAC program is off the ground. Melzer contended that WAC program websites are much more than conduits of information. Instead, according to Melzer, "WAC websites can be used to persuade, connect, and support students and faculty to create . . . a 'knowledge web': a socially constructed clearinghouse of information connected by hyperlinks in an ever-expanding web" (pp. 23–24). Melzer also goes on to say that an effective WAC program's website is not a supplement to that program, but "a center of the WAC knowledge web" (p. 24).

The field of professional communication has for some time now thought of organizational websites as not only conduits through which organizations deliver information to their audiences, but as two-way interactive communication networks that help to build relationships and communities. In 2010, before I became WAC Director at UCF, I coordinated the redesign of the website of the Association of Teachers of Technical Writing (ATTW, 2011). The ATTW had wanted a more social and interactive web platform that would not only deliver relevant information to its members but would also integrate such communication tools as forums, listservs, and so on, allowing the members to communicate with the organization and among themselves. This site redesign was done as part of a larger project conducted by the Communication Subcommittee of ATTW's Executive Committee, designed to evaluate the effectiveness of communications between the organization and its members and to improve that effectiveness. As part of this project, the Communication Subcommittee conducted a survey of ATTW members, the results of which we reported in the journal *Programmatic Perspectives* in the spring of 2012 (Popham et al., 2012). In the article, we argued that, in order for successful web communication within an organization to occur, that organization's web platform needs to offer interested parties the opportunity not only to consume information but also to contribute information and to interact with others in the community. According to our findings, "reflective analysis of communication channels is healthy for the continued success of a networked program" (Popham et al., 2012, pp. 5–41).

After becoming WAC Director at UCF, I wanted to drastically increase and improve the program's web presence, which at the time was limited to a couple of static pages on the department's website. Making regular updates on the site was difficult and took a long time because every update request had to go through the IT of the college. Such limited web presence could, of course, be explained by the newness of the WAC program itself and by the very limited scope of its activities at the time.

I wanted the new WAC program's website to be dynamic and social and to allow for content posting by multiple users. As the administrator of the future site, I also wanted the ability to change and update the site on my own and without going through the college's technical services. Given my experience with the open-source software Drupal (I also used Drupal for the ATTW website redesign and many other projects), I wanted UCF's new WAC website to be Drupal-based. Drupal is a very robust and flexible platform, which allows for the construction of highly social websites.

Because of my previous experience with Drupal-based design, in September of 2011, I was able to convince the IT services of our college to host a Drupal website for the WAC program and to give me administrative control over the design and updating of the site. I spent the fall semester of 2011 designing and informally testing the site with my colleagues in the department. The new site was rolled out in January of 2012 during a university-wide writing showcase. It contains not only the usual pages filled with information about the WAC program and its initiatives, but also forums, a mailing list, a blog, and some limited integration with social media. The rollout of these features was somewhat slow and incremental, and currently, the two most-used social features are the mailing list, through which we send out program updates to interested faculty, and our Facebook page.

Creating a truly thriving online community and an interactive web presence takes much more than a redesign of the organization's website. In order for a virtual community to function and in order for the organization's web presence to be truly interactive, the face-to-face community of that organization must thrive as well. Therefore, putting your website on a more interactive web platform will not accomplish all those goals automatically, but without such a redesign, all those steps would be impossible.

Therefore, the work of further developing the WAC program's Web presence during the 2011–2012 academic year and in future years will consist not only of filling the site up with up-to-date program information but also of creating a truly social Web presence. As the face-to-face WAC community at UCF increases and develops, so will the virtual community on the website. My immediate plans for the website's development include starting and keeping a regular blog and finding ways to encourage interested faculty members to share teaching tips and resources on the forums.

## BUILDING AND MAINTAINING PARTNERSHIPS

As I have said throughout this chapter, WAC work is the work of building and maintaining partnerships. Such partnerships need to be built not only with other academic departments, whose faculty the WAC program serves, but also within your own department and with campus-wide units and organizations, which have an impact on faculty development and instruction. In this segment of the chapter, I'd like to highlight one department-wide and one university-wide partnership that the WAC program works to maintain and develop.

The first partnership, which I consider very important to the success of WAC at UCF, is with the Writing Outreach Program in our home department—the Department of Writing and Rhetoric (DWR). The WAC program is actually a part of Writing Outreach, which also includes the first-year composition program and the University Writing Center. According to DWR's website, a part of the department's mission is to "serve UCF students, faculty, and community partners in fostering transferrable disciplinary and professional writing expertise" (Department of Writing and Rhetoric, 2012). The website further states that these learning opportunities are supported by the first-year composition program, the University Writing Center, and the WAC program. The Writing Outreach Program is more of a concept than a formal structure, but having this concept allows constituents both inside and outside the department to better understand what we do.

Having an operational concept like the Writing Outreach Program allows the WAC program to easily explain its mission and methods to various groups in the university more easily. When I give presentations to faculty and administrators about our program, I always tell them that the WAC program is a part of a three-way effort to improve writing instruction at UCF: the first-year program provides direct instruction in academic writing to students; the Writing Center supports students through individual consultations; and the WAC program supports faculty in helping them teach writing better in the courses in their disciplines.

One example of partnering between WAC and First Year Composition thus far is the participation of the WAC program in UCF's annual Knights Write showcase. Knights Write is a conference-style event during which select student panelists read from their recent writing and compete for writing awards. During the 2011–2012 academic year, the showcase included, for the first time, a WAC poster display, with students from Nursing and History demonstrating their research and writing from courses in their disciplines.

The WAC also partners with the University Writing Center in the implementation of an embedded-tutor program. As I mentioned earlier, the pilot embedded-tutor initiative had to be put on hold, but there is a considerable interest among faculty in various departments in restarting it soon. The Writing Center Director and I are working on plans for identifying prospective tutors, training them within the Writing Center educational structures, and placing them in courses in other disciplines.

To date, WAC's most important university-wide partnership is that with the UCF's Faculty Center for Teaching and Learning (FCTL). The center is charged with faculty development across the university, and like many similar bodies at other institutions, it works to achieve its goals by offering regular faculty workshops, conferences, and other forms of support.

Our partnership with FCTL this year has ranged from me serving on FCTL's Advisory Board to participating in the winter and summer faculty conferences with presentations and workshops about WAC to using FCTL's conference room for WAC Fellows Program meetings and other events. I have also written two articles about WAC for FCTL's university-wide faculty newsletter.

Our collaboration with FCTL is mutually beneficial. The WAC program is charged with faculty development through providing writing instruction assistance to faculty university-wide. Our mission, then, aligns very well with the mission of FCTL, which is also faculty development. FCTL's ability to reach many faculty members in various departments gives the WAC program a useful platform for reaching potential partners. At the same time, by partnering with FCTL, the WAC program offers the Center an opportunity to expand its faculty-development repertoire and its reach.

Certainly the partnerships with Writing Outreach Programs and with FCTL are not the only ones we'd like to pursue in the future. During the 2011–2012 academic year, we had initial contacts with the Office of Undergraduate Research as well as with several interdisciplinary NSF-funded science and technology undergraduate education initiatives on campus that would like to include scientific writing instruction in their curricula. We would also like to develop a partnership with the university Information Fluency Program.

As WAC Director, I see sustaining the existing partnerships and developing new ones as necessary conditions for the success of the WAC efforts at UCF. Being a successful consultant to a variety of clients presupposes the ability to network with those clients and demonstrate to them that you have valuable knowledge, skills, and ideas that will help them achieve their goals and objectives. One of the goals of a successful WAC program should be to position itself at the university as a source of expert advice about writing, which understands the needs and circumstances of others while also bringing into the equation the disciplinary expertise that we have to offer as writing specialists. In that sense, this work is very similar to the work of communication professionals, both in academia and in the workplace. Professional communicators are often the people who negotiate and connect various people and groups by using their expertise in writing and communication.

## LESSONS, CONCLUSIONS, PLANS

The WAC program at UCF is very new, so it is certainly too early to speak of any kind of permanent and sustained success of the program. However, during the time period described in this chapter, the program has formed some interesting partnerships on campus and achieved some promising results.

Through the WAC Fellows Program, we have impacted seven academic departments and programs by helping their faculty develop writing-related learning outcomes for their students and the assignments to support those outcomes. At present, the WAC program is implementing a variety of projects with 12 departments representing eight colleges of UCF.

Although these projects vary in scope and scale, ranging from department-level curricular revisions to consultations with individual faculty members on assignment and course design, they all follow the consultant and department-centered models of WAC. We have carefully positioned ourselves as disciplinary experts in writing who are ready to assist other faculty on campus in the development of methods and approaches to writing instruction that, first and foremost, make sense for them, their students, their departments, and their academic disciplines and professions.

Our current partners come from various disciplinary communities: humanities, business, science, technology, engineering, math (STEM), health sciences, and others. Details of implementation of various projects with these different groups of academic disciplines are dictated by their needs, structures of their curricula, requirements of their accrediting bodies, and so on. In some cases, faculty members are skeptical about our claims that only they can teach disciplinary writing to their students. At the same time, so far, we have encountered nothing but consensus on the importance of writing for the students' professional, civic, and personal growth. This campus-wide consensus helps us work on achieving our goals.

The understanding of the importance of writing for any field and profession comes not only from the grassroots: in deciding to create and fund the WAC program, the university administration has shown its commitment to the improvement of writing instruction at UCF. The kind of financial resources and support the program has enjoyed during its first years has allowed it to develop the partnerships described here. Without those resources and that support, none of this work would have been possible. Of course, at the time of across-the-board budget cuts, especially here in Florida, we need to ensure the continued financial health of the program. In addition to securing funding from the university, we are partnering with the University Writing Center, the department, and the college to seek out external funding sources through grants and collaborative ventures with industry and business. Seeking out such external partnerships is yet another example of how WAC links up with professional writing, where industry-academic connections are almost the norm.

Next year, and for years to come, we plan to work with departments on the development of writing-related learning outcomes by working with groups of faculty through the WAC Fellows program. Simultaneously, we plan to develop implementation and assessment plans with those departments whose faculty participated in the WAC Fellows project during the 2011–2012 academic year.

Simultaneous with continuing the WAC Fellows initiative, we will continue to conduct individual consultations with faculty on WAC issues. Last year's experience shows us that some of the faculty who came to us for individual consultations found them so effective and useful that they assembled teams of colleagues from their departments who became WAC Fellows.

We also plan to continue expanding our program's web presence by more actively developing the interactive tools of our website as well as our social media presence. At an institution of UCF's size, getting your website noticed and used is a challenge. However, we would like the cadre of faculty actively working with us on WAC projects to use the website not only as an information resource but also as an online platform for exchanging ideas, assignments, and so on. One immediate and usable resource that is published on the website and that can help us further expand our influence on campus are the learning outcomes and assignments developed by the WAC Fellows. In recent months, we have suggested those resources to faculty members who are new to WAC and who came to us for consultations.

Finally, we plan to continue developing our departmental and campus-wide partnerships. Our program provides a consulting service to the university community, so we need to position ourselves as writing experts with a wide and diverse reach who are able to interact with and help clients with various needs, priorities, and challenges.

## REFERENCES

Anson, C. (2012, June 8). *The department-centered model of WAC.* The International Conference on Writing Across the Curriculum, Savannah, GA.

Artemeva, N. (1998). The writing consultant as cultural interpreter: Bridging cultural perspectives on the genre of the periodic engineering report. *Technical Communication Quarterly, 7*(3), 285–299.

Association of Teachers of Technical Writing (ATTW). (2011, July 30). *Association of teachers of technical writing.* Retrieved July 18, 2012 from http://www.attw.org/

Bean, J. (2011). *Engaging ideas: The professor's guide to integrating writing, critical thinking, and learning in the classroom.* San Francisco, CA: Jossey-Bass.

Carter, M. F., Ferzli, M., & Wiebe, E. N. (2007). Writing to learn by learning to write in the disciplines. *Journal of Business and Technical Communication, 21*(3), 278–302.

Conklin, J. (2007). From the structure of text to the dynamic of teams: The changing nature of technical communication practice. *Technical Communication, 54*(2), 210–231.

Department of Writing and Rhetoric. (2012, January). *About the department.* Retrieved July 15, 2012 from http://writingandrhetoric.cah.ucf.edu/

Institutional Knowledge Management. (2012, January 1). *UCF 2011–2012 facts at a glance*. Retrieved July 19, 2012 from http://www.iroffice.ucf.edu/character/current.html

Melzer, D. (2006). WAC websites as knowledge webs. *The WAC Journal,* 23–32.

Moore, P. A., & Kreth, M. (2005). From wordsmith to communication strategist: Heresthetic and political maneuvering in technical communication. *Technical Communication, 52,* 302–322.

Popham, S., Arduser, L., Cargile Cook, K., Rush Hovde, M., Koerber, A., Mogull, S. A., et al. (2012). Evaluating the effectiveness of a network of communication channels: Lessons learned from a communication survey. *Programmatic Perspectives, 4*(1), 5–41.

Russell, D. (2007). Rethinking the articulation between business and technical communication and writing in the disciplines. *Journal of Business and Technical Communication, 21*(3), 248–377.

UCF WAC Program. (2011, December 15). *About the WAC fellows program*. Retrieved July 20, 2012 from http://wac.cah.ucf.edu/aboutfellows

Walvoord, B. (1992). Getting started. In M. S. Soven (Ed.), *Writing across the curriculum: A guide to developing programs* (pp. 9–22). Newbury Park, CA: Sage.

CHAPTER 12

# Curricular Challenges of Emphasis Degrees in Technical and Professional Communication

***Lisa Meloncon***

Undergraduate degrees in technical and professional communication (TPC) have a long history that dates back to 1958 when Carnegie Mellon University began its Bachelor of Science in Technical Writing and Editing. TPC has long understood that "it is important as a field to generate and control our own data, beginning at the program and department level" (Phelps & Ackerman, 2010, pp. 206–207), and TPC first attempted to quantify and track the number of programs it offered in 1976. At that time, early leaders identified 12 undergraduate degree programs (Pearsall & Sullivan, 1976). Current research has identified 178 undergraduate degree programs (Meloncon, 2012; Meloncon & Henschel, 2013).

When Harner and Rich (2005) reported undergraduate curricular information, they used *program* to include "majors, minors, concentrations, emphases, tracks, and specializations" (p. 210). Even though this amalgamation of all undergraduate degree programs into a single kind has a scholarly history (e.g., Balzhiser & McLeod, 2010; Cunningham & Harris, 1994; Stewart, 1989), TPC's growth rate makes it possible—and necessary—to add more specificity to curricular discussions because degree programs, especially at the undergraduate level, have important differences. Of the 178 degree programs, 65 are degrees *in TPC*. These degrees are no different from other majors found on campus, such as communication, biology, or engineering, and they are distinguished as such in college catalogs. The remaining 113 degree programs are what I refer to as *emphasis degrees* (see Anderson, 1978), which are major degree programs that include an emphasis/track/ concentration/specialization in TPC.

Because the majority of TPC undergraduate degree programs are emphasis degrees, program administrators need to have a better understanding of the challenges and benefits of these degree programs. In what follows, I weave together the local narrative of the University of Cincinnati's undergraduate emphasis degree with national, fieldwide data to illustrate ways to make the curriculum sustainable, to discuss the importance of being good neighbors within the department, to highlight the role of technology in programs, to raise difficult questions about contingent faculty and teacher preparedness, and to discuss the preprofessional status.

## OVERVIEW OF EMPHASIS DEGREES

From Harner and Rich's study in 2005 to mid-year 2012, emphasis degrees have experienced a growth rate of 86%. The majority of the emphasis degrees in my dataset were started during this time frame.[1] Basic characteristics of emphasis degrees are

- 88% are Bachelor of Arts (BA)
- 85% are administered out of English/literature departments
- 58% are found at Carnegie-classified master's-level institutions; 24% at research (extensive and intensive) universities (http://classifications.carnegiefoundation.org/)
- 44% require 12–18 hours in the track; 15 hours is the most popular
- 54% require a cumulative experience such as a capstone course or a portfolio of work
- 61% require a basic course, which introduces students to the genres of TPC; this course also serves double duty as being the "service course" for other departments
- 49% require an internship

Unlike undergraduate degrees in TPC, emphasis degrees do not have a set of "core courses" consistent across curricula nationwide. Other than the basic course, the capstone course and an internship are the only required courses that can be considered a "core course" (Meloncon, 2009, 2012). Thus, emphasis

[1] The methods for data collection in previous scholarship are incomplete, which makes it difficult to determine true growth rates. Moreover, the distinction between degrees in TPC and emphasis degrees is normally not even considered. Harner and Rich, in their 2005 study, were the first to make clear that they were combining data. Other datasets, e.g., Cunningham & Harris 1994, imply that they combined all undergraduate degrees together. Thus, although I have gone to great lengths to create a clear and open methodology (see Meloncon, 2009 and Meloncon, 2012 for additional information on the larger study) that can be replicated in the future, I am referencing past studies in which the data is not as precise.

degrees are still widely divergent and are more similar to undergraduate certificates (Meloncon, 2012), which is not surprising because most certificates also require a limited number of courses. Even with diverse curricula, there are some commonalities in the types of courses offered. For example, courses in editing (38%), specific genres (33%), web (31%), and design (29%) appear with some frequency in curricula and could be classified as "burgeoning core courses" (Meloncon, 2012, p. 213). Moreover, the wide range of English studies is also represented in required courses: creative writing (16%), linguistics (16%), journalism (17%), theory (13%), and rhetoric (13%).

The majority of emphasis degrees are similar to the description offered by Paul Anderson (1978), the first published discussion of emphasis degrees. At the 1978 CPTSC conference, Anderson reported on the "emphasis program" started at Miami University that year. More recent case studies about emphasis degrees can be found in the literature (Baker & Henning, 2010; Franke, Reid, & DiRenzo, 2010; Peeples, Rosinski, & Strickland, 2007), but what is most striking about Anderson's description is how well it has held up and more importantly, how easily it moves from the local to global, fieldwide concerns. Understanding that his audience consisted mostly of administrators of full-fledged degree programs, Anderson focused his presentation on the four general characteristics of an emphasis degree: more generalized in their objectives, more literature courses, fewer specialized courses, and fewer courses outside English.

These four characteristics materialize in emphasis degrees in quite specific ways. For example, to establish the emphasis, departments must make compromises that follow a similar pattern. In addition to the emphasis courses, students are required to take a set of "English major core courses" that typically include an introductory course to literary criticism, a Shakespeare course, a multicultural literature course, and a survey course (or two) in British and American literature. Thus, the English major core is where one finds the "more literature courses" required in an emphasis degree. (However, 20% of emphasis degrees do not require literature courses, but most of these emphasis degree programs are housed outside of English departments.)

Beyond the "English major core," emphasis degrees then have students take a set of courses that comprise the emphasis. Emphasis degrees are more generalized because they have fewer courses to work with than full-fledged TPC degree programs. Providing a foundation, breadth and depth is not easy (some might say impossible) to do in a limited number of courses. Thus, most emphasis degrees are more generalized, focusing their curricula on courses that provide transferable knowledge across genres and situations while also offering students the opportunity to explore the field and its many divergent paths. When compared to TPC degrees, emphasis degree course offerings still reflect what Anderson (1978) describes as "few specialized courses," courses with specialized content created specifically for the emphasis.

In TPC degree programs, some 58% of degree programs require students to take courses outside the department (Meloncon & Henschel, 2013, p. 54). This requirement is in place to ensure that students are receiving subject matter expertise that they can then couple with their writing experience. In contrast, as Anderson (1978) notes, emphasis degrees have fewer courses outside English. Only 11% of the emphasis degree programs require courses outside the department.

## RHETORIC AND PROFESSIONAL WRITING TRACK AT THE UNIVERSITY OF CINCINNATI

TPC has a long history at the University of Cincinnati (UC). The department currently offers an undergraduate certificate started in 1984, a master's degree started in 1989, a graduate certificate started in 1989, and a track in Rhetoric and Professional Writing (RPW), as part of the English major, started in 2008.

The graduate program had worked for many years as an almost autonomous body within the department. The proposal of the RPW track meant a tighter integration with the department. From the beginning of discussions of the RPW track, faculty were aware that they needed to create a curriculum that would allow professional writing to coexist with the traditional English department. This particular challenge is not unusual for TPC programs (Henze, Sharer, & Tovey, 2010). The first attempt to get the RPW track approved failed (which occurred 2 years before I was hired at UC in 2006), which meant that on the second try, faculty needed to "locate the best curricular framework" (Peeples, 2002) for the track that would also be recognized and accepted by the English literature and creative writing faculty.

We used a theory/practice curricular framework (Peeples, 2002) in which practice incorporates Miller's (1989) use of practice as praxis and conduct. In addition, we grounded our curriculum in scholarship that supported TPC as a "field that combined techniques, skills, and pragmatics with theory, reflection, and critique" and also emphasized the "social, interpretative dimensions of communication" (Rentz, Debs, & Meloncon, 2010, p. 286). In more practical terms, the professional writing faculty has also been adept at updating the curricula based on trends in the professional field (such as online documentation and more recently an increase in instructional design) while keeping a close eye on local markets. For example, there is not a great local demand for traditional technical writing positions associated with technology or software. Instead, local technical writers work in health, medicine, engineering, and environmental-type positions.

Initially conceived as a "curricular space for the overlapping interests of the professional writing and the composition faculty," the RPW track also signaled a "significant milestone" to "grow the intellectual and social capital of the program within with the department" (Rentz et al., 2010, p. 289). Like the discussion above, UC's emphasis degree requires students to take a series of "English major core courses," courses in the RPW track, and electives. The RPW track was conceived and implemented in much the same way that Anderson (1978) (and others) have described, that is, by looking at current course offerings to determine whether they have adequate coverage to make the argument for the emphasis. Initially, the only new course created for the RPW track was a visual rhetoric course I created when I came to UC, and thus this "specialized course" had more to do with my arrival than with diversifying the curriculum.

In the first year of RPW track, we graduated three students who managed to meet all the requirements for the track. Those first graduates are presently working in a library, completing law school, and working as a writing design specialist for a national marketing firm headquartered in Cincinnati. The diversity of these first three graduates illustrates, much like Anderson outlined in 1978, how emphasis degrees are more general in objectives because they prepare students for a broad range of careers (p. 50).

Building on our previous curricular narrative of surviving and thriving in an English department (Rentz et al., 2010), the following discussion of UC's emphasis degree outlines our considerations, as we had to rethink our emphasis curriculum in light of "semester conversion." As the track was entering it's second year in 2009, UC launched semester conversion, which was a result of a state legislative mandate that universities on the quarter system would convert to the semester system. For those unfamiliar with the quarter system, degree programs under quarters require more credit hours. Instead of two semesters and the summer, we had three 10-week quarters and the summer (the summer quarter is considered the fourth quarter). Students would typically take six or seven courses a quarter, and with the increased number of hours for degree, students could take a wider variety (or breadth) of courses. At UC, semester conversion was a 3-year process with an effective date of fall 2012. So at a time when faculty should be evaluating a new curriculum, we had institutional permission to assess and evaluate what we wanted our emphasis degree to do under semesters (see Appendix A for the quarter curriculum and Appendix B for the semester curriculum).

Keeping in mind the curricular framework that had served the professional writing faculty, the department, and the students so well for almost 30 years, we made three important changes to the RPW track under semesters. First, we made the course, *Introduction to Rhetoric and Professional Writing*, a required course for all students in the track. This course offers a common foundation for understanding, theorizing, and applying rhetorical principles and for developing advanced writing skills. It also introduces students to professional writing as an academic and professional field. Although some have advocated for an introductory course in a writing major (Balzhiser & McLeod, 2010), only 16% of TPC emphasis degrees require an introductory course. A key feature of our course is a series of guest speakers who work as professional writers in a wide range of positions. One recent term featured a health communicator working for a government agency, a Web designer, an e-learning specialist, a social media manager, and a technical writer. This course also introduces the e-portfolio students are required to complete in the program. Moreover, with recent changes to staffing (see below), we felt we could consistently offer this course, thus mitigating a constant administrative headache.

Second, we maintained the requirement of a specialized writing course and a capstone course. Genre-based curricular frameworks are not uncommon in TPC programs, requiring students to take a series of courses that focus predominately on exposure to a wide variety of genres (such as reports, brochures, web content, etc.) with attention to audience and purpose concerns. We felt a course that focused on writing genres related to business, technical, biomedical, or environmental fields afforded students the opportunity to produce a variety of workplace genres as well as practicing such skills as collaboration and visual communication. These specialized writing courses also play other key institutional roles that contribute to curricular sustainability, discussed in more detail later in this chapter.

We also maintained the requirement of having students take the capstone course. Of the emphasis degree programs that require a capstone experience, 42% of them require a course. The capstone is a semester-long course that brings together the curriculum in a series of client-based projects. Our requirements for a client generally include a web component; a substantial, genre-based component (such as a report); and several smaller tasks. A recent term included restructuring the navigation of a website, rewriting and writing all of the content for the website, creating a series of one-page information sheets, producing a minute-long video public service announcement, creating two brochures, and writing and researching a feasibility report.

Third, we created a series of additional courses that would be rotated to give students ample opportunities to explore the diversity and breadth of professional writing. These additional courses also make it possible for students to take several classes suited to their own particular interests. For example, if a student is interested in visual communication, she or he can select a series of courses in that area. Because we now offer more courses that cover more of the breadth of the field, we no longer have to rely on courses in journalism or creative writing to fill out the curriculum (though students can still take these courses with their general English electives or general education electives). These additional courses not only match the faculty expertise, they also intersect with current trends in the professional field. For example, we now offer a data visualization course that picks up on the themes of "big data" and the role a technical and professional communicator can play in designing and delivering this information. We also created additional courses that intersect with concepts and trends in composition that also maintain the integrity of a professional writing program. Examples of these courses include Multimedia Writing and Rhetoric and Public Engagement. The creation of these specialized courses moves the UC emphasis degree away from Anderson's (1978) characteristics and also away from many other programs.

This rethinking and implementation of the emphasis degree for semesters would not have been possible without additional faculty. The RPW track was approved in my second year at UC (in 2007), and in the same year, we were able to secure a renewal of a full-time, non–tenure track line. In 2008, we hired a tenure-track faculty member in composition, and in 2010, through university realignments, we added three additional faculty members (another tenured faculty and two full-time, non–tenure track positions). This hiring meant we could offer required and a range of elective courses more regularly ensuring students could graduate in a timely fashion.

Although not a radical reconfiguration of the emphasis degree, these three changes provide more coherence so that students can plot a trajectory through courses that have been systematically designed and offered. Using semester conversion as a way to integrate our curriculum means students now have the same background, thus making subsequent courses easier to structure. Semester conversion also afforded us the opportunity to ensure that we were all using the same terminology for the same things. Perhaps most importantly, the changes done under semester conversion were possible because we had already done the hard work of instituting the track, and we were departmentally aware enough to recognize that the only latitude we had under semester conversion was with the track and not the English major core. There is a constant negotiation with other parts of English studies in English departments, and we are aware of this relationship in every move that we make.

## IMPLICATIONS FOR TPC PROGRAM ADMINISTRATORS

There are numerous lessons for other program administrators in the UC story. I highlight five of the most important: creating a sustainable curriculum, being good neighbors, the role of technology, impacts on labor, and preprofessional status. Appendix C provides a checklist, of sorts, which contains a series of questions program administrators and faculty should consider when planning a new emphasis degree or updating an existing one.

### Creating a Sustainable Curriculum

Stevenson (1978) created a map for "purposeful exploration of your own technical writing course offering" (p. 203). His map, which may be obvious to anyone fluent in the scholarship of teaching and learning and curriculum design, reminds anyone involved in program planning and upkeep not to "lose sight of curriculum relationships of any substantive sort" (p. 205). Programmatic outcomes and individual course outcomes are necessary for any sustainable curriculum (see Bridgeford & Ilyasova, this volume). This attention to outcomes when planning emphasis degrees is especially important because emphasis degrees must take into consideration the outcomes of the major (e.g., English) and the emphasis. When we approached semester conversion, we relied heavily on outcomes in deciding which new courses to create. Faculty expertise in an area did not necessarily warrant immediate inclusion of a new course if that course could not be envisioned and created to meet the overall programmatic outcomes. Rather, we focused on ensuring that students had a solid grounding in rhetorical approaches and/or communication theories, critical thinking and reading, knowledge of design, technological proficiency, and writing in myriad genres (see Appendix D for UC's RPW outcomes). For a sustainable curriculum, a routine evaluation of courses is encouraged to maintain a clear sense of what the program is trying to do and to understand where the weaknesses may lie so that program administrators and faculty are ready to act if resources or circumstances become available.

In addition to focusing on outcomes, thinking through ways to get "double duty" out of courses can also help sustain emphasis degrees. The specialized courses that we created are crossover courses, which are the courses that serve advanced undergraduate students and graduate students. UC is not unusual in having a series of courses that serve dual roles. Crossover courses are a necessity for many emphasis degree programs with limited faculty and growing numbers of students. Crossover courses provide necessary room to provide curricular diversity and to afford room for programs to grow at multiple levels until arguments can be made for an increase in faculty lines or other resources. Even with our seeming bounty of new faculty, our program could not be sustained without these courses. Additionally, crossover courses have also become a way to offer more specialized courses in topics that students will recognize. Said another way, they not only serve a specific role in the curricula but they also serve as marketing tools. For example, students had long taken an editing course in journalism because the course was available, frequently offered, and for the longest time, we did not have the faculty to staff the course. Under semesters, we now offer our own editing course, which is drawing our professional writing students (both undergraduate and graduate) as well as students from other areas. However, crossover courses do not come without a price. They are notoriously hard to teach, even harder to assess, and completely absent from research in TPC. In programmatic literature, one can find several references to, and concerns about, differences between undergraduate and graduate programs (Keene, 1997, p. 195), and these issues of quality and degree are even more pressing when considered in light of crossover courses. Even with the inherent problems, these courses are usually necessary for program sustainability.

One of the most time-consuming and contentious discussions about changing the curriculum for semester conversion had to do with the number of courses to offer every term and a balance of required and elective courses. As program administrators know, nothing will stifle the enthusiasm for a major like required courses not being offered in a timely fashion. As described above, we decided to require *Introduction to Rhetoric and Professional Writing,* which means that the course must be offered regularly. It only takes around 45 students in the track for the need to offer the course every term, thus starting a staffing snowball effect. If someone has to teach *Introduction to Rhetoric and Professional Writing* every term, then one less elective or "service course" course is offered. To help ease the workload burden, we were strategic in maintaining the requirement that students take a specialized writing course (e.g., Writing for Business and Technical and Scientific Writing). Because we offer these courses regularly as service courses, we were able to provide majors with a necessary course in practicing writing for specific audiences while not needing to offer additional courses.

We were able to update and keep the entire track within the department without sacrificing coverage and topics. This move allowed us to ensure the coherence of the emphasis and have greater control of course offerings, which we felt was a key to long-term sustainability. However, smaller institutions (where most emphasis degrees are housed) may not be able to make this move. In these cases, program administrators may need to "look beyond our discipline for the teaching of media production, multimedia design, and information management" (Reid, 2010, p. 254) and form partnerships with other faculty to ensure coverage of topics. These models can be just as sustainable, but require a different kind of work, such as relationship building and being meta-aware of changes within those departments offering courses for the emphasis: "From a sustainable perspective, we must constantly remind ourselves that we are part of the problem *and* the solution" (Johnson, 2004, p. 113). We must balance our desire for emphasis degrees with the resources available to us. In the current environment in higher education, full-time faculty lines are a precious resource, and the sustainability of most programs is directly tied to having faculty to teach these courses. Sustainable curricula must take advantage of every resource within and outside of the department, and program administration must be mindful of innovative ways to deliver a coherent set of courses that can provide value to students when they enter the workplace.

## Being Good Neighbors

In our programmatic narrative about how to survive and thrive in an English department (Rentz et al., 2010), we were explicit in stating that to feel a part of an English department, "the professional writing faculty must become visible contributors in the department and find ways to integrate their program's purposes into the department's priorities" (Rentz et al., 2010, p. 290). Admittedly, for me, this claim has always rankled. In some ways, it means that professional writing faculty have to work just a little bit harder to keep up with other areas of English studies without expectation of any reciprocation. Even though it rankled then (and still rankles now), it is not an untrue statement.

Emphasis degrees are overwhelmingly administered out of English Departments (88%). Concerns about the English "core curriculum" courses are ones that program administrators must be particularly attentive to. The *Introduction to English Studies* course, the one that was lauded in the history of program (Rentz et al., 2010, p. 294) was receiving a lot of push-back from faculty. For the most part, the course was being taught by junior literature faculty who felt they did not have the expertise to adequately incorporate all areas of English studies. The situation was further exacerbated by other internal issues about course assignments that included the fact that none of the TPC faculty were available to teach the course because of curricular needs at the graduate level. Thus, the *Introduction to English Studies* course is now, once again, an *Introduction to Literary Criticism* course under the semester curriculum. This was a change that the TPC faculty endorsed.

We recognized that not only was this course a battle we did not need to fight, but this change also gave us leverage to create our own introductory course. Thus, the TPC faculty did not feel as though we were losing anything substantial in the shift. In fact, by readily agreeing to the change, our literature colleagues were able to see us—once again—as willing colleagues and true collaborators in the undergraduate curriculum. We were also reminded to maintain attentiveness to these larger departmental concerns and to pick battles wisely and cautiously.

## The Role of Technology

As I was collecting data for my large curricular history project, I had to make some important methodological decisions. Much curricular scholarship in the past has relied on self-reported data. Self-reporting can be a viable means of data collection, but it also is fraught with many problems, not the least of which is a self-reported dataset that may not be representative of the field. I wanted to mitigate this problem, so I opted to gather the data myself and then have it interrated by other TPC specialists, as well as a non-TPC specialist, to increase the reliability and validity of the dataset. For my project, interrating was primarily used to determine what programs could "qualify" as a TPC program and which ones would not. Often programs were excluded not because of what courses they offered but rather by what courses they did not offer.

> The field of technical communication is in many ways inscribed by technology. As a result, technical communication programs not only must provide students with a foundation in the theory and practice of the field, but also must give students some level of proficiency in the technology tools they will need to put that knowledge into service in the workplace. (Brumberger, 2003, p. 64)

Thus, many programs, even if they called their emphasis degree "professional writing" (or some other variation of the name), were "excluded" from the dataset because the emphasis degree program did not include courses with a substantial technology or technological component. Programs need to think long and hard about the relationship their curriculum has with technology and how to ensure that issues of technology (both critique and production) are included in the curriculum.

With rapid changes in technology, keeping abreast of the latest software and trends can be a daunting task. Thus, there is an ongoing need for professional development for faculty as well as ensuring the program has enough faculty for coverage of the technology-centric courses. At UC, because I am the "technical" person, I teach the majority of the "technology courses," which can cause ripple effects if for some reason I am not available to teach my full course load. For example, I was recently named the director of a new research center on campus, which means I do not teach my full course load. We are reconsidering and adjusting the 2-year curriculum plan to ensure that the required technology courses are offered. Second, program administrators have to be aware that the specialized software used in industry are complex, multilayered tools that require in-depth study and practice. It is difficult to teach a course that integrates a specialized software if the instructor has never used the software. Thus, administrators have to develop sustainable ways to provide faculty professional development opportunities to stay abreast of technological changes.

Technology issues play a larger role in TPC programs because they are pivotal and vital to the work we do. Beyond basic digital writing technologies and understanding the impact of new media on delivery of information, TPC programs are also faced with questions about specialized software used throughout the professional field of technical communication. Routinely professional listservs will discuss favorite tools or make note of the "must know" tools for technical communicators. In technical communication pedagogy, there is an ongoing tension between teaching tools (i.e., actual and specific commercial software packages) and teaching how to use tools. Students are often insistent on learning Program X because they see it repeatedly listed in job ads, and Lanier's (2009) examination of job ads provides tangible evidence of this point. On the other hand, programs often promote the development of technological literacy so that students can quickly and easily learn the fundamentals of any software program. With rapid changes in technology and ongoing increases in costs, tool debates are likely to grow in frequency and pitch. Beyond programmatic outcomes, TPC program administrators face an uphill battle concerning the role of technology and courses that incorporate specialized software into the curriculum. From a programmatic standpoint, these sorts of courses can be expensive. Thus, strategic decisions and long-range (generally 5 years) plans must be made if the program will be paying for or sharing the cost of a computer lab.

Though programs promote a broader approach to technology, they still work hard at providing lab space and access to many of the industry standard software, such as Adobe's Creative Suite, which includes Photoshop, InDesign, Dreamweaver, Flash, Acrobat, and several other programs. Industry standard software brings two distinct challenges. First, these tools are quite expensive. Even with educational discounts, software for one computer can run a program about $1,500. (My colleagues and I calculated this figure in the spring of 2010 when we had to officially submit a proposal to maintain our professional writing lab space. In the face of budget cuts, administration planned on eliminating all lab spaces, but they did provide programs with specific needs to argue for continuance. Our proposal and presentation saved our lab space for 3 years.) Should programs choose to require students to purchase the software as they would a course textbook or to add a "technology fee" to registration are big questions that require careful consideration based on local situations. These discussions need to occur to ensure that the best decisions are being made for the programs' sustainability and meeting the curricular needs. When coupled with the hardware costs, limited educational spaces, ongoing maintenance and upgrades, negotiating for technology and lab spaces becomes an ongoing struggle.

## Contingent Faculty

Although TPC has never defined itself through its service obligations, there is no escaping the fact that it carries a large service load. Most institutions with any type of TPC degree also administers the service courses, which are "introductory courses for nonmajors delivered primarily as a service to other departments and programs on campus" (Meloncon & England, 2011, p. 398): "The service course is a particularly rich site of contention, promise, and crisis—a bellwether of the field" (Knievel, 2007, p. 89), and as such, discussions of the service course are particularly important within larger discussions of emphasis degrees (and other degree programs). In times of tight budgets, there is a direct cause and effect between the creation of new degree programs and the service course. The result of this interaction is usually the increased reliance on contingent faculty.

In discussions of creating new writing majors in composition, Tony Scott (2007) called for caution without due consideration of the "institutional pressures, constraints, and professional contradictions that already characterize writing education" (p. 82). Scott's emphasis on the material realities created by the institution's bureaucracy was written as a cautionary tale for those in composition who wanted to create new majors. But what is most striking are the parallels between Scott's argument and the realities of "the material terms of labor that frame everyday writing pedagogy" (p. 83) that are already being played out at institutions with TPC programs and at institutions offering just the service course in TPC.

In the first study to examine contingent labor in TPC, Meloncon and England (2011) found that 83% of all service courses were taught by non–tenure line faculty. This holds true at my institution. With semester conversion curricular changes and the growth in the number of students selecting the RPW track, in AY 2012–2013, tenure-line faculty will teach only one section of the service course. Full-time non–tenure track faculty will teach half of the sections, with the remaining sections being taught by adjuncts and graduate students. This cause/effect within the UC program is indicative of Scott's concerns and also indicative of what is happening at other institutions when the move is made to start degree programs. Even with the increased staffing described previously, we are, at the time of writing this chapter, at capacity. If the RPW track and undergraduate certificate continue to grow, we will have to move our educator faculty into RPW track courses, thus increasing the contingent labor problem at the service course level.

Closely related to this question of contingent labor is faculty preparedness. Additional problems and considerations include the necessity for increased professional development opportunities for faculty teaching new courses and for the contingent faculty teaching the service courses. Space issues preclude me from delving into this issue of preparedness at great length, but the bigger question facing the field—particularly in light of the growing number of emphasis degrees—is what makes someone qualified to teach in a TPC program? Is it a degree in composition? Workplace experience? A declaration that the faculty member self-identifies as a TPC scholar/teacher? A combination of all three or something entirely different? These larger questions directly intersect and impact discussions of contingent faculty. As programs rely more heavily on contingent faculty discussions, these discussions have to take place, and they are discussions the field has not taken up in any sustained way.

## Preprofessional Status

Thomas Miller (2011) suggested that more accessible institutions, such as baccalaureate- and master's-level institutions have more TPC degree programs because they often feel greater pressure to create vocational degrees (p. 234). Even though some have argued against the "preprofessional" label (Gordon, 2009) and the professional status of technical communicators continues to be a source of academic debate (Coppola, 2011, 2012; Kynell-Hunt & Savage, 2003, 2004; Pringle & Williams, 2005), TPC cannot escape the fact that our programs are in effect creating a preprofessional, someone who is about to enter a professional career. However, "there is important middle ground between selling out to corporate America and providing critical instruction in efficacious knowledge. Our students will find it extremely difficult to survive economically if we fail to prepare them to communicate in contemporary workplace" (Henze et al., 2010, p. 79).

What this statement means is that TPC programs must adapt and change their curricula more often because of changes in business and industry and of advances in technology. One of the biggest outcomes of our change in the curriculum for semesters was an increased integration of technology use and technology critique into many of our courses. Even though the process of writing has not altered much, the way writing is composed and delivered has altered dramatically. Take for example the idea of online documentation. Products and services still provide online help, but the way they provide that help has shifted to include user interaction and multiple delivery modes. Programs must continually adapt or risk being labeled as outdated and out of touch, which is already a constant refrain made by practitioners toward academic programs.

## LOOKING TO THE FUTURE

As Robert Connors' (1982) historical work has shown, there has always been a tension in the relationship between industry practice and academic instruction of TPC. Ongoing debates in higher education surround big questions like what is the purpose of higher education? How can job training be balanced with a liberal arts education? How can programs continually do more with less in the current economic climate? And myriad others are all questions that TPC program administrators already know. TPC has always sat in an interdisciplinary middle ground with an eye toward the workplace. Thus, program administrators and faculty need the "the perseverance to sustain our programs, and to anchor our power and status in and out of the academy, will require deep and sometimes painful assessment of who we are and where we want to go" (Johnson, 2004, p. 113). These "painful assessments" are difficult in the best of circumstances, but emphasis degrees—more so than other degree progams—require this sort of attention because with so few courses to "certify" that students have a specific set of skills or a range of knowledge, there is an imperative that the curriculum has a clear direction and focus.

A constant theme found in the pages of the CPTSC proceedings is one of collaboration and sharing of information so that TPC administrators can design successful programs. To do this, "faculty need to know what such programs look like" (Anderson, 1978, p. 58) and the challenges they may face. Having models that can be adapted to local situations is one of the best ways to ensure that TPC emphasis degrees provide value to students, faculty, departments, institutions, and the communities our institutions serve.

# APPENDIX A
## Quarter RPW Track Curriculum

UNIVERSITY OF Cincinnati

## English Major – Rhetoric and Professional Writing Track

This track prepares students to write persuasively and effectively in public and professional contexts. Students in this track may design a program of study that leads to employment in commercial and non-profit organizations, to graduate work in rhetoric and professional writing, and/or to enhanced literacy and more informed, responsible participation in civic affairs.

### Requirements

- A minimum of 54 credit hours is required, with at least 18 hours in the chosen track.
- Only courses listed on this checklist count toward the major. If you wish to substitute other courses, you must submit a Petition to Substitute a Course before taking the course.

| **Core Courses for all English majors – 24 Hours** | Term/Year | Grade |
|---|---|---|
| ○ 15ENGL300 – Intro to English Studies | ______ | ______ |
| ○ 15ENGL305 – Writing for English Majors | ______ | ______ |
| ○ 15ENGB306 – Intro to Shakespeare | ______ | ______ |
| Survey: American Literature – choose 2 courses for a total of 6 hours: 15ENGL301, 15ENGL302, OR 15ENGL303 | | |
| ○ ______ | ______ | ______ |
| ○ ______ | ______ | ______ |
| Survey: English Literature – choose 2 courses for a total of 6 hours: 15ENGB334, 15ENGB335, OR 15ENGB336 | | |
| ○ ______ | ______ | ______ |
| ○ ______ | ______ | ______ |
| Multicultural Literatures in English – choose 1 course for a total of 3 hours: *(see reverse side for a list of recommended courses)* | | |
| ○ ______ | ______ | ______ |
| **Required R&PW Track Course – Choose 1 course for a total of 3 hours** 15ENGL359, 15ENGL491, OR 15ENGL492 | | |
| ○ ______ | ______ | ______ |
| **Additional R&PW Track Courses – Choose 4 courses for a total of 12 hours** *(see reverse side for a list of approved courses)* | | |
| ○ ______ | ______ | ______ |
| ○ ______ | ______ | ______ |
| ○ ______ | ______ | ______ |
| ○ ______ | ______ | ______ |
| **Capstone Course – 3 hours** **Prerequisite of 9 hours in R&PW Track Courses* | | |
| ○ 15ENGL589 – Rhetoric and Professional Writing Capstone | ______ | ______ |
| **English Electives – Choose 4 English courses, at the 200 level or above, for a total of 12 hours** | | |
| ○ ______ | ______ | ______ |
| ○ ______ | ______ | ______ |
| ○ ______ | ______ | ______ |
| ○ ______ | ______ | ______ |

Effective 11S

UNIVERSITY OF Cincinnati

**English Major – Rhetoric and Professional Writing Track**
**Requirements Page 2**

## Multicultural Literatures in English Courses

**See adviser for approval of courses not on this list**

15ENGC342 – Third World Literature
15ENGC363 – African Literature I: Intro
15ENGC364 – African Literature II: Topics
15ENGC365 – African Literature III: Drama
15ENGC366 – Caribbean Literature
15ENGC373 – Modern Fiction III
15ENGC504/505/506 – Intro to Folklore
15ENGC581 – Exile in Displacement
15ENGL341/342/343 – American Ethnic Literature
15ENGL344 – Interracialism in America
15ENGL382 – Lesbian and Gay Lit
15ENGL388 – American Women Writers: pre-1900
15ENGL389 – American Women Writers: post-1900
15AFAM310/311/312 – African American Lit

## Additional R&PW Track Courses

15ENGL359 – Intro to Rhetoric and Writing
15ENGL360 – Discourse Communities
15ENGL370 – Desktop Publishing
15ENGL376 – Writing with Style
15ENGL377 – Visual Rhetoric
15ENGL368 – Topics in Rhetoric and Prof Writing
15ENGL489 – Advanced Composition
15ENGL491 – Writing for Business
15ENGL492 – Technical and Scientific Writing
15ENGL499/500/501 (3-hr. maximum) – Internship
*(Prereq: One 600-level Writing Sem or perm. of inst.)*
15ENGL580 – Writing Cincinnati
15ENGL621 – Prof Writ Sem: Instructional Design
*(Prereq: 15ENGL491, 15ENGL492, or perm. of inst.)*
15ENGL622 – Prof Writ Sem: Promotional Writing
*(Prereq: 15ENGL491, 15ENGL492, or perm. of inst.)*
15ENGL623 – Prof Writ Sem: Reports & Proposals
*(Prereq: 15ENGL491, 15ENGL492, or perm. of inst.)*

One JOUR course from the list below may be used as an Additional R&PW Track Course. A prerequisite of 15JOUR214 and 15JOUR215 will apply, and these 200-level courses do *not* count toward the major.

15JOUR301 – Columns and Reviews
15JOUR302 – Feature Writing
15JOUR303 – Copy Editing
15JOUR426 – Magazine Writing
15JOUR525 – Narrative Nonfiction
15JOUR622 – Editing Principles and Practices
15JOUR624/625/626 – Journalism Seminar

## English Major – R&PW Track Curriculum Planning Guide

The model schedule below is a guide for planning only. Majors should see their English major adviser early to discuss major requirements and plan their schedules.

**First Year – 42-54 hours**
- English Composition* (6)
- Foreign Language * (9-15)
- Natural Science* (9-15)
- History* (9)
- Social/Behavioral Sciences* (9)

**Second Year – 51 hours**
- 15ENGL300 (3) and 15ENGL305 (3)
- Multi-cultural Literatures course (3)
- Two Survey of American Lit courses (6)
- English Composition* (3)
- Mathematics* (9)
- Foreign Language (if necessary)* (9)
- Natural Science (if necessary)* (9)
- Social/Behavioral Sciences* (6)

**Third Year – 48 hours**
- Two Survey of English Lit courses (6)
- 15ENGB306 (3)
- 15ENGL389, 15ENGL491, OR 15ENGL492 (3)
- Additional R&PW Track courses (6)
- English electives (6)
- Free electives* (15)
- Humanities* (9)

**Fourth Year – approx. 45 hours**
- Additional R&PW Track courses (6)
- English electives (6)
- Capstone course (3)
- Free electives as needed*

*Needed to fulfill A&S College Requirements as outlined in the A&S Bulletin. Approved course lists are available in 102 McMicken Hall or www.artsci.uc.edu.

Effective 11S

## APPENDIX B
## Semester RPW track Curriculum

The English Major requires **36 semester hours with at least 15 semester hours in the RPW track**. Only courses on the approved course list can be used. See back for approved courses.

| English Major Core = 15 Hours | Term | Hours | Grade |
|---|---|---|---|
| ENGL 3000 Introduction to Literary Studies | | | |
| ENGB 3006 Shakespeare | | | |
| ENGL 2007 or 2008 American Literature Survey Course | | | |
| ENGB 2034 or 2035 British Literature Survey Course | | | |
| Ethnic and Minority Literature Course (see back for list) | | | |
| **RPW Track Core = 9 hours** | | | |
| ENGL 3059 Introduction to RPW | | | |
| ENGL 4000 Specialized Writing Course | | | |
| ENGL 5095 Capstone (9 hrs. RPW courses as prereq.) | | | |
| **RPW Electives = 6 hours** | | | |
| | | | |
| | | | |
| **English Electives = 6 hours** | | | |
| | | | |
| | | | |
| | | | |
| **Sem. Hours = _____**<br>**Total Semester Hours Needed = 36** | | | |

## Approved Course List

**4000 Specialized Writing Courses**

ENGL 4091 Writing For Business
ENGL 4092 Technical and Scientific Writing
ENGL 4093 Environmental Writing
ENGL 4094 Biomedical Writing

**Note: These courses are in high demand and fill quickly. Contact your advisor or the PW Program Director should you have problems registering.**

**RPW Courses for Electives**

ENGL 2073 Rhetoric and Social Engagement
ENGL 3060 Lang. & Knowledge Discourse
ENGL 3067 Research Methods in RPW
ENGL 3068 Topics in RPW
ENGL 3069 Introduction to Web Authoring
ENGL 3070 Desktop Publishing
ENGL 3072 Multimedia Writing
ENGL 3074 Rhetoric of Social Media
ENGL 3076 Writing With Style
ENGL 3077 Visual Rhetoric
ENGL 4099 Internship #
ENGL 4095 Theory in Professional Writing
ENGL 5080 Writing Cincinnati
ENGL 5091 Individual Work
ENGL 6021 Instructional Design
ENGL 6022 Promotional Writing
ENGL 6023 Reports and Proposals
ENGL 6024 Editing Professional Documents
ENGL 6025 Info Graphics & Doc Design
ENGL 6026 User Experience and Usability
ENGL 6028 Publishing and New Media

**#Note: See advisor at least two quarters prior to enrollment. This internship must be arranged through the PW program and not Professional Practice.**

**Ethnic and Minority literature courses**

ENGL 3041 Intro to American Ethnic Lit
ENGL 3042 Topics in American Ethnic Lit
ENGL 3043 Race in African American Lit
ENGL 3044 Interracialism
ENGL 4008 Native American Women's Lit
ENGL 4009 Ethnic Women's Literature
ENGL 4035 Lesbian Literature & Psychology
ENGC 3010 Introduction to Folklore
ENGC 3011 The Folktale
ENGC 3060 Literature of the African Diaspora
ENGC 3063 African Literature
ENGC 3066 Caribbean Literature
ENGC 4010 Third World Women's Lit
ENGC 4011 African Women's Literature
ENGC 4021 Female Bildungsroman
ENGC 6081 Exile in Modern Literature

**Advising Notes:**

____________________________________________________

____________________________________________________

# APPENDIX C
## Questions to Consider in Creating or Updating an Emphasis Degree

### Institutional Questions

- Do other majors on campus have emphases?
- What required paperwork is necessary to start/revise an emphasis degree? How much buy-in is necessary?
- What local constraints will influence the configuration of the emphasis degree?

### Departmental Questions

- How will this emphasis impact the overall major (i.e., will other emphases be added at the same time or will this be only one)?
- How many hours will the emphasis degree be allowed within the major?
- What will the core English major look like?
- Does the major have a required cumulative experience (such as a capstone), and can this course be adjusted for emphasis students?
- What is the climate like for TPC? Has there been success with other TPC programs? Is there a respected service obligation?

### Emphasis-Related Questions

- What are our program outcomes?
- When students leave, what do we want them to look like?
- What guiding curricular framework will make the emphasis degree coherent rather than a random set of courses?
- What ideal courses would accomplish program outcomes?
- What courses are already created that can form the foundation of the emphasis degree? If the need exists to look outside the department for emphasis courses, what are the standard policies or procedures in place at your institution with regard to these sorts of relationships?

### Specific Features

- Are there computer labs or existing policies (such as laptop requirement) that will enable the integration of technology into all courses?
- What impact will the required technology have on students? Faculty?
- Is it possible to provide students an internship or co-op experience? If so, how will this requirement be managed and sustained?
- What role will service-learning (or client-based) projects have in the curriculum?

## Administration and Faculty Considerations

- Will the department create an administrative position (if one is not already created) to direct the emphasis? If not, how will it be administered?
- Who will be responsible for advising (i.e., TPC faculty or all faculty or a combination)?
- What impact will the emphasis have (staffing and other resources) on other parts of the department?
- Are enough faculty available to teach required courses regularly?
- What impact will this emphasis degree have on other teaching obligations (such as service commitments)?
- Will the department and/or the TPC faculty be able to offer professional development opportunities for faculty?
- Do we have faculty who are adequately prepared to staff these programs?
- How will the program be assessed?

## APPENDIX D
## RPW Outcomes

- Identify, articulate, and focus on a defined purpose for the creation of written texts
- Respond appropriately to the needs of the relevant audience
- Demonstrate an ability to write in multiple genres (using conventions and formats) appropriate to the situation
- Employ flexible strategies for generating, revising, editing, and proof reading texts
- Use appropriate technologies to organize, present, and communicate information to address a range of audiences and purposes
- Analyze the contexts and relationships among which language, information, and knowledge are produced, managed, organized, and disseminated the ways that social, cultural, historical, and economic issues relate to information, writing, and technology
- Understand the ethical dimensions of information and technology
- Incorporate design principles into documents (both print and online)
- Understand the role, importance, and methods of conducting research in professional writing
- Be competent in using a range of available technological applications, such as Adobe's Creative Suite and web authoring tools, to fulfill both written and visual needs of documents produced

## REFERENCES


Anderson, P. V. (1978). *The undergraduate English major with an emphasis in technical writing.* Paper presented at the CPTSC.

Baker, L., & Henning, T. (2010). Writing program development and disciplinary integrity: What's rhetoric got to do with it? In G. Giberson & T. A. Moriarty (Eds.), *What are we becoming: Developments in undergraduate writing majors* (pp. 153–173). Logan, UT: Utah State University Press.

Balzhiser, D., & McLeod, S. H. (2010). The undergraduate writing major: What is it? What should it be? *College Composition and Communication, 61*(3), 415–433.

Brumberger, E. (2003). *Beyond the borders of "English": Teaching technology tools in the undergraduate technical communication curriculum.* Paper presented at the Council of Programs in Technical and Scientific Communication, Potsdam, NY.

Connors, R. J. (1982). The rise of technical writing instruction in America. *Journal of Technical Writing & Communication, 12*(4), 329–352.

Coppola, N. W. (2012). Professionalization of technical communication: Zeitgeist for our age, introduction to this special issue (Part 2). *Technical Communication, 59*(1), 1–7.

Cunningham, D. H., & Harris, J. G. (1994). Undergraduate technical and professional writing programs: A question of status. *Journal of Technical Writing & Communication, 24*(2), 127–137.

Franke, D., Reid, A., & DiRenzo, A. (Eds.). (2010). *Design discourse: Composing and revising programs in technical and professional writing.* Fort Collins, CO: WAC Clearinghouse and Parlor Press.

Gordon, J. (2009). The pedagogical missions of professional and technical communication programs: What we say in the journals and what we say on the web. *Programmatic Perspectives, 1*(2), 112–138.

Harner, S., & Rich, A. (2005). Trends in undergraduate curriculum in scientific and technical communication programs. *Technical Communication, 52*(2), 209–220.

Henze, B. R., Sharer, W., & Tovey, J. (2010). Disciplinary identities: Professional writing, rhetorical studies, and rethinking "English." In D. Franke, A. Reid, & A. Di Renzo (Eds.), *Design discourse: Composing and revising programs in professional and technical writing* (pp. 63–86). Fort Collins, CO and Anderson, SC: WAC Clearinghouse and Parlor Press.

Johnson, R. (2004). (Deeply) sustainable programs, sustainable cultures, sustainable selves: Essaying growth in technical communication. In T. Kynell-Hunt & G. J. Savage (Eds.), *Power and legitimacy in technical communication Volume II: Strategies for professional status* (pp. 101–119). Amityville, NY: Baywood.

Keene, M. L. (Ed.). (1997). *Education in scientific and technical communication: Academic programs that work.* Arlington, VA: Society for Technical Communication.

Knievel, M. (2007). *Growing the service course: Anticipating problems, promise in the technical communication "mini-program."* Paper presented at the Council of Programs in Technical and Scientific Communication, Greenville, NC.

Kynell-Hunt, T., & Savage, G. J. (Eds.). (2003). *Power and legitimacy in technical communication, Volume 1: The historical and contemporary struggle for professional status.* Amityville, NY: Baywood.

Kynell-Hunt, T., & Savage, G. J. (Eds.). (2004). *Power and legitimacy in technical communication, Volume 2: Strategies for professional status.* Amityville, NY: Baywood.

Lanier, C. R. (2009). Analysis of the skills called for by technical communication employers in recruitment postings. *Technical Communication, 56*(1), 51–61.

Meloncon, L. (2009). Master's programs in technical communication: A current overview. *Technical Communication, 56*(2), 137–148.

Meloncon, L. (2012). Current overview of academic certificates in technical and professional communication in the United States. *Technical Communication, 59*(3), 207–222.

Meloncon, L., & England, P. (2011). The current status of contingent faculty in technical and professional communication. *College English, 73*(4), 396–408.

Meloncon, L., & Henschel, S. (2013). Current state of US undergraduate degree programs in technical and professional communication. *Technical Communication, 60*(1), 45–64.

Miller, C. R. (1989). What's practical about technical writing? In B. E. Fearing & W. K. Sparrow (Eds.), *Technical writing: Theory and practice* (pp. 14–24). New York, NY: Modern Language Association.

Miller, T. P. (2011). *The evolution of college English: Literacy studies from the Puritans to the postmoderns*. Pittsburgh, PA: University of Pittsburgh Press.

Pearsall, T. E., & Sullivan, F. (Eds.). (1976). *Academic programs in technical communication* (1st ed.). Arlington, VA: Society of Technical Communication.

Peeples, T. (2002). *A call for prax-ical undergraduate major curricular framework.* Paper presented at the Council of Programs in Technical and Scientific Communication, Logan, UT.

Peeples, T., Rosinski, P., & Strickland, M. (2007). Chronos and kairos, strategies and tactics: The case of constructing Elon University's professional writing and rhetoric concentration. *Composition Studies, 35*(1), 57–76.

Phelps, L. W., & Ackerman, J. M. (2010). Making the case for disciplinarity in rhetoric, composition, and writing studies: The visibility project. *College Composition and Communication, 62*(1), 180–215.

Pringle, K., & Williams, S. D. (2005). The future is the past: Has technical communication arrived as a profession? *Technical Communication, 52*(3), 361–370.

Reid, A. (2010 ). The write brain: Professional writing in the post-knowledge economy. In D. J. Frank, A. Reid, & A. Di Renzo (Eds.), *Design discourse: Composing and revising programs in professional and technical writing* (pp. 254–273). Logan, UT and Anderson, SC: WAC Clearinghouse and Parlor Press.

Rentz, K., Debs, M. B., & Meloncon, L. (2010). Getting an invitation to the English table and whether or not to accept it. *Technical Communication Quarterly, 19*(3), 281–299.

Scott, T. (2007). The cart, the horse, and the road they are driving down: Thinking ecologically about a new writing major. *Composition Studies, 35*(1), 81–93.

Stevenson, D. (1978). Mapping the unexplored area: Developing new courses and coherent programs in technical communication. *Journal of Technical Writing & Communication, 8*(3), 193–208.

Stewart, D. C. (1989). What is an English major, and what should it be? *College Composition and Communication, 40*(2), 188–202.

CHAPTER 13

# Afterword: 40 Years of Stories from Technical Communication Program Administrators

***Karla Saari Kitalong***

As Tracy and Bill wrote in the introduction, this collection gathers narrative reflections written by members of the professional, technical, and scientific communication (PTSC) community who have served or who now serve as academic program administrators. The value of stories is undisputed in our culture, as Kirk St. Amant explains in his Foreword to this collection. Stories also structure the discipline of technical communication program administration. A case in point: For much of its 40-year existence, the Council for Programs in Technical and Scientific Communication (CPTSC) has relied on stories—lore—as the framework for its annual meeting. At each CPTSC meeting, participants present short position papers, which are published as program abstracts and later as conference proceedings. Many of these papers are based on or inspired by the characters, settings, and plots that structure program administrators' work days. I'll wager that for most attendees, the lengthy discussions that follow each presentation, during which more stories are exchanged and commented upon, constitute not just the framework, but the real "meat" of the conference. Ephemeral though these interactions are, they continue to be widely touted as the reason people return to the conference year after year. Through such discussions with their fellow administrators, in short, attendees are inspired to "restory" (Dubinsky, this volume) by brainstorming, characterizing, and sometimes solving their intellectual and institutional challenges.

The Council has certainly come of age as a professional organization since its initial meeting in Minneapolis in 1972. Healthy professional organizations often provide vehicles for publishing the scholarship of the discipline; CPTSC does this through its journal, *Programmatic Perspectives*, and the online publication of its conference proceedings. Another characteristic of thriving organizations is a means of materializing ephemeral artifacts; for CPTSC, a robust online presence allows institutional memories—meeting websites, e-mail list postings, job ads—to be collected, digitized, and shared.

Thriving organizations also document their own histories. CPTSC has just begun to do this; its archives, for example, are held at the University of Minnesota, where the first meeting of the organization was convened in 1972. Working from these archives, Tracy Bridgeford presented an entertaining and instructive multimedia historical overview of the organization, titled "Around the Table," at the conference in Minneapolis in 2008.

*Sharing Our Intellectual Traces* solidifies several previously ephemeral narratives. In so doing, it preserves the storytelling impetus of CPTSC while adding context and theorizing the stories in such a way as to inspire new and experienced administrators alike to examine their roles as program leaders in a rapidly changing educational landscape. As I comment on the pieces included in this edited collection, I want simultaneously to engage with the history of PTSC program administration by reflecting on a revealing but little known and rarely cited historical resource, a series of 20 articles published between the mid-1970s through the 1980s in the *ADE Bulletin*, the journal of the Association of Departments of English. At the time, technical communication was an emerging academic specialty. Program administration had not yet been validated as a scholarly endeavor, and the scholarship of administration was published only sporadically (if at all) in the English studies journals. Nevertheless, department chairs needed help understanding this mysterious new world that was neither literature nor creative writing, and not even really composition (although composition, also still in its formative stages, was its nearest relative). These articles, even though they didn't necessarily cite one another and they have not been widely cited in the intervening years, fulfilled a need and remain an important historical snapshot of an emerging field.

Together with the chapters in this volume, they bookend the 40-year history of PTSC program administration. The *ADE Bulletin* articles published from the mid-1970s to the mid-1980s were engaged in identifying the key moments, trends, and questions that would ultimately come to shape the field. From their vantage point in the second decade of the 21st century, recent CPTSC presidents Deborah Andrews, Steve Bernhardt, Kelli Cargile Cook, Jeff Grabill, Bruce Maylath, Dan Riordan, and Stuart Selber (this volume) provide a state-of-the-discipline report by asking themselves, "Of the many currents that flow together to create our professional contexts, on which should we focus? What are the key trends, particularly in the areas of pedagogy, technology, globalization, budgets, and the job market?" And, I would add, for the purposes of this afterword, what has changed in the intervening years?

Many issues alluded to in these long-ago articles remain salient today. Then, as now, the academia/workplace divide needed to be bridged, both by job-seeking English majors and by the faculty members who had been enlisted to teach technical communication. Effectively teaching workplace communication practices to students required the literature faculty who found themselves in this role to assume an unfamiliar mindset (Babinsky, 1985; Barnum, 1983; Bickley, 1974; Driggers, 1986; Merrill, 1989; Norman, 1986; Raymond, 1982; Turnbull, 1981).[1] Today we understand that we have much to learn from our colleagues in business (Morgan, this volume), but back in the 1970s and 1980s, it was a revelation to English department chairs that the expansion of scientific and technical work into the business world provided an impetus for incorporating technical content and style into humanities disciplines (Crow, 1985; Whitburn, 1975). Whitburn (1984) later outlined some differences between academic and business contexts, including "security precautions [that] prevent the release of proprietary information (the information that gives a corporation its competitive edge)" and the fact that "professors have greater freedom than do industrial employees . . . to pursue knowledge that promotes public or private ends." On the other hand, the "value universities place on publication" benefits both business and academia in that "research results will be made available to others as quickly as possible." Then as now, academics had an incentive to work with businesses: even in 1984, the "gradual decline of federal support for research" was forcing "university professors and universities to turn to the corporate world for help" (Whitburn, 1984).

Other questions that occupied the chairs' attention back then, and that we still entertain today, include how best to design technical communication programs that are compatible with the institutional context and that address needs of potential employers (Pearsall, 1982). Certainly Andrews et al. (this volume) address the problem of program administrators not fully understanding how our own institutions operate. "We want to encourage teachers and administrators to study how their institutions plan, develop, support, and manage" resources; they speak specifically of technology, but it makes sense to engage in this sort of mapping across the board so as to develop the kind of "big picture view" that is "crucial to creating strong, sustainable frameworks that can support program activity." Pearsall (1982) suggested the following strategies for forming a technical communication program at a time when there were 25 institutions of higher learning offering some form of technical communication coursework, and a growing market for technical writers.

[1] I located these articles by searching CompPile (www.comppile.org) and then locating specific references in the ADE's online archive (www.ade.org). Because the articles are in HTML format, pagination does not match the original, therefore I do not include page numbers in the parenthetical citations.

The first question, according to Pearsall (1982), was "Are there people on your staff or in related departments who can teach such subjects as graphics, oral communication, interviewing, and communication theory?" Pearsall added, "Course work in science and technology should be available," as well, an unfamiliar and perhaps uncomfortable expectation for English departments. He also argued that a program's success depends upon a receptive departmental staff, one that does not "see the program as mere camouflage for teaching more literature." Pearsall further advised that the "community environment is also important," and recommended that program directors consider how the institution's proximity (or not) to an urban area might affect program administrators' ability to find "practicing technical writers" to serve on an "advisory committee," and offer "internships and, ultimately, employment" for the program's graduates.

Pearsall's (1982) main objective was to urge department chairs to "plan new programs intelligently, based on the knowledge that already exists about what technical communicators need." He admonished against a "hastily put together collection of writing and editing courses," concluding that a "program needs the proper environment to flourish" (Pearsall, 1982). Similarly, Andrews et al. (this volume) suggest that program administrators consider all the distinct ways to understand the field and situate individual programs within their unique social and intellectual contexts.

Several authors discussed the cultural divide between the humanities and the technical disciplines (Buchen, 1982; Zentner, 1981); for Crow (1985), "The typical businessperson's stereotype of a humanist closely resembles Rodin's thinker—preoccupied, inert, and fun to parody." Rivers (1980) questioned whether business and technical writing could be considered "humanistic" at all, given the reliance on templates and formats. He resolved the question, in a sense, by 1985, when his survey of department chairs' approaches to staffing revealed that most business and technical communication courses were taught by faculty who had traditional literary training, occasionally augmented by specialized technical communication training (Rivers, 1985); overwhelmingly, technical communication teachers were products of a humanities education.

S. Michael Halloran (1984) offered a primer on technical communication scholarship for department chairs. In so doing, he distinguished between "technical communication" and "technical writing," raising a question about programmatic nomenclature that still surfaces to this day. Moreover, he diminished the gulf between technical communication and traditional English studies disciplines by noting that technical communication scholarship "shares assumptions, methods, and goals with the traditional concerns of the English department and thus has a place in English studies." Further relating the field and its scholarship to the larger field of English studies, he defined technical communication as "applying the principles of rhetoric to problems of communicating within and beyond specialized fields of discourse."

Notwithstanding its "important affinities with literary study and composition research," Halloran (1984) states that technical communication scholarship has "equally important differences" from literary studies, including "some research methods that are foreign to the literary scholar." This remains true today: Usability studies in technical communication, for example, employ laboratory-based research methods such as think-aloud protocol, eye-tracking, and contextual inquiry. Another difference, according to Halloran, is that technical communication "views a text as a tool rather than as an artifact, as a means toward some end rather than as a thing of intrinsic value." Redish's (1989) taxonomy of how people read—to learn, to do, and to learn to do—exemplifies this particular difference between literary studies and technical communication. Finally, wrote Halloran, technical communication scholarship "takes into account the means by which a text will be produced and the medium in which it will be presented to an audience." Lisa Meloncon and Julie Dyke Ford's chapters (this volume) demonstrate that these classic curricular planning questions are still relevant.

In short, Halloran (1984) reminded department chairs of the affinities between technical communication and literary studies by noting that technical communication scholarship not only "can benefit from the critical insights of literary scholarship," but also can "contribute to a fuller understanding of the processes and products of literacy." Coney, Ramey, and Souther (1984), then as now associated with a prominent technical communication program situated in an engineering department, noted that a program's success hinges more on the "quality of the staff and the support it receives than on the organizational placement of the program." Having said that, however, they went on to specify certain conditions that must be in place for technical communication programs to succeed, regardless of where they are placed in the institution. For one thing, technical communication teachers must be "grant[ed] academic equality within the department." Moreover, technical communication should not be "isolated from the world of work"; students and teachers should not be alienated from each other; and the program should eschew a "limited view of writing." At that time, an expanded view included instruction and immersion in the publication process and in editing. Today, a broad-based program includes not only written and oral communication, but also visual media design, usability studies, and networked digital communication. Crow (1985) would appreciate an expanded view of the field: "good writing consists of clarity and of correctness in grammar and spelling. Certainly a well-prepared English major can deliver more than this."

Some details of what students should be expected to know and be able to do as a result of having graduated from a technical communication program—such things as software abilities and knowledge about the theory and practice of usability testing—have changed over the years. Yet much remains the same; for instance, Corbett (1984) advocated an approach to technical communication in which first-year students concentrate on honing their critical reading skills, while intensive writing courses are offered to students in their junior or senior years. This approach resonates with the writing across the curriculum movement, the popularity of which has ebbed and flowed over the past 40 years; in fact, a recent article in my university's daily faculty/staff newsletter touts a resurgence of writing across the curriculum on our campus, which was recognized as one of the earliest adopters of writing-intensive curricula in the country. In their chapters in this volume, Pavel Zemlianski and Julie Dyke Ford each outline strategies for working across the curriculum to develop and strengthen existing technical communication efforts on their respective (and very different) campuses.

Similarly, some program administrators' duty areas that are missing from the earlier articles—namely, technology management, assessment, and international technical communication—have become serious endeavors for today's technical communication program administrators. In this volume, technology theory and user-centered design practice are integrated into a narrative about the revitalizing of a computer lab (Brady and Kitalong); assessment is addressed through two pieces: Nancy Coppola's narrative about her growing passion for assessment and Bridgeford and Ilyasova's proposal for a standard set of technical communication outcomes similar to the ones distributed by the Council of Writing Program Administrators. Laurence José thoughtfully situates technical communication programs, faculty, and students within the so-called global context, a term that rolls easily off of the tongue despite its inherent complexity. And Ritu Raju discusses the challenges of offering a technical communication program in a 2-year college, where research issues seldom penetrate.

Certainly there's much more this collection could collect, and there's much more I could say to draw connections between the *ADE Bulletin* articles and the chapters gathered in this volume. My original title, which I used when I talked about the *ADE* articles at a previous CPTSC conference, was, "The More Things Change . . . Issues in Technical Communication Program Administration from the *ADE Journal*, 1974–1985." I prefer the title I ended up with for this collection, because it retains the primacy of storying—and restorying—as valued practice in PTSC program administration. I hope you enjoyed the collection as much as Tracy, Bill, and I have enjoyed bringing it to you.

## REFERENCES

Babinski, H. F. (1985). From academe to the business world. *ADE Bulletin, 81,* 36–38.

Barnum, C. M. (1983). English professors as technical writers: Experience is the best teacher. *ADE Bulletin, 76,* 32–34.

Bickley, R. B., Jr. (1974). Career opportunities for English majors in the Southeast: A Florida perspective. *ADE Bulletin, 40,* 38–40.

Buchen, I. H. (1982). Business and the humanities: A new cultural detente? *ADE Bulletin, 72,* 32–33.

Coney, M. B., Ramey, J. A., & Souther, J. W. (1984). Technical writing in the English department: An outsider perspective. *ADE Bulletin, 79,* 40–42.

Corbett, E. P. J. (1984). A collegiate writing program for the 1980s. *ADE Bulletin, 78,* 20–23.

Crow, P. (1985). An English major for the information age: Language resources for management. *ADE Bulletin, 81*, 13–15.

Driggers, S. (1986). An instructor internship in technical writing. *ADE Bulletin, 83,* 46–48.

Garvey, S. (1983). Internships for English majors: A new dimension for the humanities. *ADE Bulletin, 75,* 13–15.

Halloran, S. M. (1984). What every department chair should know about scholarship in technical communication. *ADE Bulletin, 79,* 43–45.

Joyce, J. J., & Sutherland, A. C. (1981). Rhetoric, student writing, and the marketplace. *ADE Bulletin, 69,* 35–38.

Merrill, L. L. (1989). Notes from the other side: The strange profession of technical writing. *ADE Bulletin, 92,* 25–29.

Norman, R. (1986). Careers in technical writing: Advising English majors. *ADE Bulletin, 84,* 30–34.

Pearsall, T. E. (1982). Building a technical communication program. *ADE Bulletin, 71,* 15–17.

Raymond, M. W. (1982). Enhancing the connections between English and the non-academic world. *ADE Bulletin, 72*, 27–31.

Redish, J. C. (1989). Reading to learn to do. *IEEE Transactions on Professional Communication, 32,* 289–293.

Rivers, W. E. (1980). The place of business writing in English departments: A justification. *ADE Bulletin, 65,* 27–31.

Rivers, W. E. (1985). The current status of business and technical writing courses in English departments. *ADE Bulletin, 82,* 50–54.

Tryzna, T. (1983). Careers for English majors: Where are they and how can departments help? *ADE Bulletin, 75,* 9–12.

Turnbull, A. D. (1981). Technical writing in the computer industry: Job opportunities for Ph.D.'s. *ADE Bulletin, 67,* 26–30.

Walzer, A. E. (1979). Business writing and the liberal arts tradition. *ADE Bulletin, 61,* 23–27.

Whitburn, M. D. (1975). Technical communication: An unexplored area for English. *ADE Bulletin, 45*, 11–14.

Whitburn, M. D. (1984). Freedom in the research and teaching of rhetoric: University-industry cooperation. *ADE Bulletin, 79,* 37–39.

Zentner, R. D. (1981). Business and the humanities: A new look at two cultures. *ADE Bulletin, 68,* 13–16.

# Contributors

**Deborah C. Andrews** is professor of English and directs the Center for Material Culture Studies at the University of Delaware. In the English department, she teaches courses in technical editing and rhetoric, in interpreting objects and sites for public understanding, and in American literature from 1865 to 1945. Through her minor in material culture studies, she supervises internships, coordinates a yearly colloquium series, and teaches a seminar in strategies for researching and writing about the relationship between people and their things. She has published several articles, book chapters, and texts on professional communication, including *Technical Communication in the Global Community* and, most recently, *Management Communication: A Guide.* A researcher, consultant, and speaker on many aspects of professional communication, especially in an international context, she is the former editor of *Business Communication Quarterly.*

**Stephen A. Bernhardt** holds the Andrew B. Kirkpatrick Jr. Chair in Writing at the University of Delaware, from which position he promotes strong writing and communication skills across the university. He is the author of *Writer's Help*, a new, web-based reference handbook from Bedford/St. Martin's. He teaches courses in scientific and technical communication, first-year composition, computers and writing, and grammar and style.

**M. Ann Brady** is associate professor of rhetoric and technical communication and director of the Scientific and Technical Communication Program at Michigan Technological University. Working at the intersections of feminist theory, rhetoric, technology studies, and interdisciplinary studies, she has published in the *Journal of Business and Technical Communication*, *Technical Communication*, *Technical Communication Quarterly*, *Women's Studies*, *Rhetoric Review*, and *Programmatic Perspectives*.

**Tracy Bridgeford** is an associate professor of Technical Communication at the University of Nebraska at Omaha where she also directs the Technical Communication program and the English master's program. She contributed a chapter to *Resources in Technical Communication: Outcomes and Apparoaches, Teaching Writing with Computers: An Inroduction*, and *Innovative Approaches to Teaching Technical Communication*, which she also co-edited. She has also published in *Kairos: A Journal of Rhetoric, Technology, and Pedagogy*. She co-edited a special issue of *Technical Communication Quarterly* on *Techne* (2002). She is co-editor of *Programmatic Perspectives*, journal of the Council for Program in Technical and Scientific Communication and serves on its executive committee as Information Officer.

**Kelli Cargile Cook** is associate professor of technical communication and rhetoric at Texas Tech University. She has served as president of both the Association of Teachers of Technical Writing (2009–2011) and the Council for Programs in Technical and Scientific Communication (2006–2008). She served as undergraduate and doctoral program director at Utah State. She currently coordinates and directs service course instruction in technical communication at Texas Tech. Her research focuses on online education in technical communication, but she also studies technical communication pedagogy, program development, and program assessment.

**Nancy W. Coppola** is founding director of the Master of Science Program in Professional and Technical Communication at New Jersey Institute of Technology. Her publications have twice won the best research paper award from the IEEE Professional Communication Society and the APEX 2012 award for publication excellence. Her teaching has been recognized by the STC Jay R. Gould Award for excellence in teaching technical communication. She is a fellow of the STC and senior member of IEEE. Her research and publication focuses on professionalization issues for technical communication, including a body of knowledge and programmatic issues including program assessment.

**James M. (Jim) Dubinsky** is director of undergraduate studies in the Department of English and associate professor of rhetoric and writing at Virginia Tech. From 1998 until 2007 he was the founding director of the Professional Writing Program in the Department of English, and from 2008 to 2011 he served as the founding director of the Virginia Tech Center for Student Engagement and Community Partnerships, now VT-Engage. In addition, he has served as an officer on the Council for Programs in Technical and Scientific Communication and ABC, and he is currently serving as ABC's executive director. Dubinsky has received college-level awards for teaching and outreach and two university awards for the scholarship of teaching and learning. His research interests include the scholarship of engagement, the scholarship of teaching, and the rhetoric of citizenship.

**Julie Dyke Ford** is associate professor of mechanical engineering and technical communication at New Mexico Tech. She is an award-winning teacher with more than 15 years of experience teaching technical communication and working with STEM faculty to promote writing across the curriculum. Her research interests include technical communication pedagogy, communication within engineering, and knowledge transfer. Her work has been published in the *Journal of Technical Writing and Communication*, *Technical Communication Quarterly*, *Technical Communication*, the *Journal of Engineering Education*, the *Journal of STEM Education*, *IEEE Transactions on Professional Communication*, and *Composition Forum*. She has held leadership roles and served as a long-standing member of the Association of Teachers of Technical Writing and the Council for Programs in Technical and Scientific Communication.

**Jeff Grabill** is professor of rhetoric and professional writing and chair of the Department of Writing, Rhetoric, and American Cultures at Michigan State University. He is a senior researcher with WIDE Research (Writing in Digital Environments). He is also a cofounder of Drawbridge LLC, an educational technology company. As a researcher, Grabill studies how digital writing is associated with citizenship, learning, and knowledge work practices. He has published two books on community literacy and articles in journals such as *College Composition and Communication*, *Technical Communication Quarterly*, *Computers and Composition*, and *English Education*.

**K. Alex Ilyasova** is assistant professor and director of the Professional and Technical Writing Program at the University of Colorado at Colorado Springs. Her research interests include writing program administration, with a focus on technical communication programs, and identity and literacy studies. She has most recently published an editorial in *Programmatic Perspectives* (March 2012), "An Editorial Introduction of a Curriculum Showcase." Additionally, she has a forthcoming co-authored book chapter on the experience of working as the lone faculty/administrator of professional, technical, and scientific communication programs. Her previous publications include book chapters on this issue of visibility and identity in LGBT studies and queer business practices.

**Laurence José** works as assistant professor of writing at Grand Valley State University, Michigan. She earned her PhD in rhetoric and technical communication from Michigan Technological University in 2010. She also holds a Diploma of Advanced Studies with an emphasis in linguistics from the University Marc Bloch (France). Her teaching experience includes courses in technical communication, professional writing, business communication, composition, French, and linguistics. Her latest research explores the meaning of the global context for the disciplinary construction of technical communication. Her other research interests include international and intercultural technical communication, technical communication pedagogy, literacy studies, and linguistics.

**Karla Saari Kitalong** is an associate professor of Humanities at Michigan Technological University, where she also serves as Director of Composition and the Michigan Technological Multiliteracies Center. Her research interests include visual rhetoric and usability in technical communication, especially concerning new media contexts; multimodal composition pedagogy; writing program administration; and writing in the disciplines.

**Bruce Maylath** is professor and director of graduate studies in English at North Dakota State University, where he teaches courses in linguistics and technical communication. His current research takes up translation issues in technical communication and has been published in *IEEE-Transactions in Professional Communication*, the *Journal of Business and Technical Communication*, and *Technical Communication Quarterly*, among others. Along with Dale

Sullivan and Russel Hirst, he is the co-editor of *Revisiting the Past through Rhetorics of Memory and Amnesia.*

**Lisa Meloncon** is associate professor of technical communication at the University of Cincinnati, where she teaches graduate and undergraduate courses. Her research interests examine health communication and the Internet, visual rhetoric, as well as pedagogical practices and programmatic concerns. She has published widely on these topics and is the editor of *Rhetorical AccessAbility: At the Intersection of Technical Communication and Disability Studies.*

**Meg Morgan** is associate professor of English, working at the University of North Carolina, Charlotte, since fall 1987. From 1994 to 2003 she administered the first-year writing program, and after that coordinated the graduate and undergraduate programs in technical/professional writing. She is a graduate of Purdue University's Rhetoric and Composition program. She was a recipient of the Bank of America Teaching Excellence Award in 2008, was faculty president from 2006 to 2007, and publishes in the discipline of technical communication practices. She has served on several national and regional committees: the Association of Teachers of Technical Writing Teaching Committee, the Executive Committee of the Carolinas Writing Programs Administrators, and the board of the national Council of Writing Program Administrators. During her time at UNC Charlotte, she has taught many different types of classes, mostly graduate and undergraduate courses in the Technical/Professional Writing program and in Rhetoric and Composition for English MA students. She has consulted on writing and writing programs nationally and internationally. Although currently semi-retired, Morgan is still teaching and serving on several MA and PhD committees.

**Dan Riordan** began teaching Technical Writing at the University of Wisconsin-Stout in 1970. During his career he taught a wide range of Technical Communication courses from Writing Manuals to Rhetoric of Technology. During his time at UW-Stout he assisted in developing a Technical Writing minor, Technical Communication B.S. and M.S. During the last 10 years of his time at UW-stout he served as the Director of UW-stout's Nakatani Teaching and Learning Center, a leading faculty development site. Riordan was recognized with a number of awards, including the Society for Technical Communication's Jay Gould Award for Excellence in teaching, Associate Fellow of STC, ATTW Fellow and CPTSC Distinguished Service Award. Riordan is currently revising his textbook, *Technical Report Writing Today* (Cengage/Wadsworth).

**Ritu Raju** is professor in the English Department at Houston Community College. She received her PhD in technical communication and rhetoric from Texas Tech University. Her work has been published in leading professional journals, and she presents at national and international conferences. She is actively engaged in several professional bodies, including the Association of Teachers of Technical Writing and the Council for Programs in Technical and Scientific Communication. Her research interests include intercultural communication, outsourcing/offshoring, cultural usability, social networking and its use in pedagogy, New Media tools in teaching composition and distance education, and writing center theory.

**Stuart S. Selber** is associate professor of English at Penn State, where he works in the rhetoric program. He is a past president of the Association of Teachers of Technical Writing and the Council for Programs in Technical and Scientific Communication, and a past chair of the 4Cs Committee on Technical Communication. His books include *Multiliteracies for a Digital Age* (Southern Illinois University Press), *Central Works in Technical Communication* (Oxford University Press), *Rhetorics and Technologies: New Directions in Writing and Communication* (University of South Carolina Press), and *Solving Problems in Technical Communication* (University of Chicago Press).

**Kirk St.Amant** is professor of technical and professional communication and of international studies at East Carolina University.

**Bill Williamson** is a professor of Professional and Technical Writing at Saginaw Valley State University (SVSU). He has administered undergraduate technical writing programs at two universities—SVSU and the University of Northern Iowa. He has served as co-editor for Programmatic Perspectives, information officer, and president for the Council for Program in Technical and Scientific Communication. His research interests include program administration; technical communication curriculum design, and technical communication pedagogy.

**Pavel Zemliansky** is associate professor in the Department of Writing and Rhetoric at the University of Central Florida. He is also the director of UCF's Writing Across the Curriculum (WAC) program. He teaches courses in professional writing, rhetoric, and composition. As WAC director, he develops and coordinates training programs and consultations for faculty across the disciplines, assisting them in the development of writing instruction frameworks and methods appropriate for their fields. His research focuses on professional communication in international contexts and writing across the curriculum.

# Index